CASE STUDIES IN FLUID MECHANICS WITH SENSITIVITIES TO GOVERNING VARIABLES

Wiley-ASME Press Series List

CASE STUDIES IN FLUID MECHANICS WITH SENSITIVITIES TO GOVERNING VARIABLES

M. Kemal Atesmen

This Work is a co-publication between ASME Press and John Wiley & Sons, Ltd.

Registered Offices
John Wiley & Sons, Inc., 111 River Street, Hoboken, NJ 07030, USA
John Wiley & Sons Ltd, The Atrium, Southern Gate, Chichester, West Sussex, PO19 8SQ, UK

Editorial Office
The Atrium, Southern Gate, Chichester, West Sussex, PO19 8SQ, UK

For details of our global editorial offices, customer services, and more information about Wiley products visit us at www.wiley.com.

Wiley also publishes its books in a variety of electronic formats and by print-on-demand. Some content that appears in standard print versions of this book may not be available in other formats.

Library of Congress Cataloging-in-Publication Data

Names: Atesmen, M. Kemal, author.
Title: Case studies in fluid mechanics with sensitivities to governing
    variables / Dr. M. Kemal Atesmen.
Description: Hoboken, NJ : Wiley, 2019. | Series: The Wiley-ASME Press series
    in mechanical engineering | Includes bibliographical references and index.
    | Identifiers: LCCN 2018038907 (print) | LCCN 2018039274 (ebook) | ISBN
    9781119524717 (Adobe PDF) | ISBN 9781119524878 (ePub) | ISBN 9781119524786
    (pbk.)
Subjects: LCSH: Fluid mechanics–Case studies.
Classification: LCC TA357.3 (ebook) | LCC TA357.3 .A84 2019 (print) | DDC
    620.1/06–dc23
LC record available at https://lccn.loc.gov/2018038907

Cover Design: Wiley
Cover Image: © matdesign24/iStockphoto

Set in 10/12pt TimesLTStd by SPi Global, Chennai, India

Printed and bound by CPI Group (UK) Ltd, Croydon, CR0 4YY

10 9 8 7 6 5 4 3 2 1

Contents

Series Preface

The Wiley-ASME Press Series in Mechanical Engineering brings together two established leaders in mechanical engineering publishing to deliver high-quality, peer-reviewed books covering topics of current interest to engineers and researchers worldwide. The series publishes across the breadth of mechanical engineering, comprising research, design and development, and manufacturing. It includes monographs, references, and course texts. Prospective topics include emerging and advanced technologies in engineering design, computer-aided design, energy conversion and resources, heat transfer, manufacturing and processing, systems and devices, renewable energy, robotics, and biotechnology.

Preface

I have been fascinated by thermo-science fields from childhood on. Simple observations such as the motions of a flying kite, bubbles rising in a boiling kettle, a leaf floating on the surface of a lake, the force you need to get the ketchup out of its bottle, and so on encouraged me to strive to understand the science behind these and many more natural phenomena that are governed by fluid mechanics, heat transfer, and mass transfer. In college and in real engineering life, I learned that knowing only the governing equations in these vast scientific fields, and studying how they evolved and learning by heart their dependent and independent variables, was only the starting point in solving the engineering problems.

Making correct assumptions while using governing equations, using the correct thermophysical properties, and applying the right experimental results to different physical conditions are crucial in approaching the solution of problems in fluid mechanics, heat transfer, and mass transfer. I gained experience in these fields as I tackled more and more problems. Even with today's high-speed computational solutions to intricate, coupled, and non-linear partial differential equations, you have to be experienced in the assumptions you make for governing equations and their boundary conditions, the thermo-physical properties you can use, and the experimental data you apply.

In this book I will provide sensitivity analyses to well-known everyday fluid mechanics, heat transfer, and mass transfer problems. Sensitivity analyses of dependent variables on independent variables provide an engineer with understanding of the crucial variables that have to be focused on during the design process and the uncertainty that has to be quantified in simulation results. The present book is an expanded version of the book I published in 2009, which only covered case studies in heat transfer [1].

In this book a wide range of 24 practical fluid mechanics, heat transfer, and mass transfer problems are investigated. The approach to problem solution starts with the applicable basic laws of fluid mechanics, heat transfer, and mass transfer. These basic laws are the conservation of matter, the conservation of momentum, the conservation of energy, and the second law of thermodynamics. Each problem solution starts with simplifying engineering assumptions and identifies the governing equations and dependent and independent variables. In some cases, where solutions to basic equations are not possible, historical experimental studies are utilized. Another critical area is the determination of the appropriate thermophysical properties of the fluid under investigation. Then, the solutions to the governing equations, alongside experimental studies, are presented graphically. These analyses are extended to the sensitivities of the dependent variables to the independent variables within the boundaries of interest. These sensitivity analyses narrow the field and range of independent variables that should be focused on during the engineering design process.

Acknowledgments

Over 40 years of engineering, engineering management, and project management in the global arena covering the automotive, computer, data communication, and offshore oil industries have been accomplished with exceptional support from my wife, Zeynep, and my family members. Some years I was away from home for more than six months, tackling challenging engineering tasks.

I would like to dedicate this book to all the engineering project team members with whom I have had the pleasure of working over the years, with thanks for their enthusiasm, imagination, and determination. Over these years they kept coming back to work with me, without reservation.

M. Kemal Atesmen
Santa Barbara, CA

About the Author

M. Kemal Atesmen completed his high school studies at Robert Academy in Istanbul, Turkey in 1961. He received his BSc degree from Case Western Reserve University, his MSc degree from Stanford University, and his PhD from Colorado State University, all in mechanical engineering. He is a life member of ASME. He initially pursued an academic and an industrial career in parallel, and became an associate professor in mechanical engineering before dedicating his professional life to international engineering management and engineering project management for 33 years. He helped many young engineers in the international arena to bridge the gap between college and professional life in the automotive, computer component, data communication, and offshore oil industries.

He has published seven books, 16 technical papers, and four patents. His books are: *Global Engineering Project Management* (CRC Press, 2008); *Everyday Heat Transfer Problems – Sensitivities to Governing Variables* (ASME Press, 2009); *Understanding the World Around through Simple Mathematics* (Infinity Publishing, 2011); *A Journey Through Life* (Wilson Printing, 2013); *Project Management Case Studies and Lessons Learned* (CRC Press, 2015); *Process Control Techniques for High Volume Production* (CRC Press, 2016); and *Engineering Management in a Global Environment: Guidelines and Procedures* (CRC Press, 2017).

Introduction

In this book a wide range of 24 practical fluid mechanics problems that include heat transfer and mass transfer are investigated. The approach to problem solutions starts with the applicable basic laws of fluid mechanics, heat transfer, and mass transfer. These basic laws are the conservation of matter, the conservation of momentum, the conservation of energy, and the second law of thermodynamics. Each problem solution starts with the simplifying engineering assumptions and identifies the governing equations and dependent and independent variables. In some cases, where solutions to basic equations are not possible, historical experimental studies are utilized. Another critical area is the determination of the appropriate thermophysical properties of the fluid under investigation. Then, the solutions to the governing equations, alongside experimental studies, are presented graphically. These analyses are extended to the sensitivities of dependent variables to independent variables within the boundaries of interest. These sensitivity analyses narrow the field and range of independent variables that should be focused on during the design process.

The development of fluid mechanics started with the well-known law of Archimedes regarding the buoyancy of submerged bodies in the third century BC. After Newton's laws of motion were introduced, Bernoulli introduced his fluid flow equations for frictionless flow under gravity forces during the eighteenth century. Fluid flow equations, including shear forces due to viscosity, were introduced in the nineteenth century by Navier–Stokes. During the same century, Reynolds differentiated between laminar and turbulent flow regimes. Reynolds named the most important dimensionless group in fluid mechanics: the Reynolds number, the ratio of inertia forces to viscous forces. At the beginning of the twentieth century, Prandtl observed the changes in flow close to a solid boundary, namely the boundary layer. Many engineers, physicists, and mathematicians tried to solve the Navier–Stokes boundary layer equations in laminar and turbulent flow regimes under every conceivable condition for internal and external flows, and tried to verify these solutions with experiments. As computers' speed and memory capacity advanced, so did the solutions to the Navier–Stokes equations and the field took a new name: "computational fluid mechanics."

Another significant portion of fluid mechanics is the thermophysical properties of fluids, such as density, viscosity, surface tension, and so on under different temperatures and pressures. Historically, these thermophysical properties for different fluids were determined experimentally as the need arose.

I chose a wide variety of simple and fun problems in this book in order to give readers an insight into different approaches to a solution in fluid mechanics, heat transfer, and mass transfer. The sensitivities of the dependent variables to the governing independent variables are investigated under appropriate physical conditions.

Chapter 1 is about draining fluid from a tank, which uses Bernoulli's equation, namely the conservation of mechanical energy along a streamline in a steady flow, along with the experimentally determined discharge velocity coefficient from the discharge hole.

Chapter 2 treats the vertical rise of a weather balloon. Several assumptions are made by neglecting the effects of wind, humidity, clouds, thermals, reduction in gravity, Coriolis forces, and so on, in order to simplify the problem. Then, Archimedes' buoyancy law is used, along with the ideal gas law, all the way through the upper stratosphere.

Chapter 3 treats the stability of a right circular cone-shaped object floating in water. Again, Archimedes' buoyancy law is used to find the tipping conditions for the cone. Similar applications can be formulated for any object floating on the surface of a fluid.

In Chapter 4 the wind drag forces acting on a person are investigated. All or some of the wind's kinetic energy is converted into pressure energy as a person tries to stagnate the oncoming wind. The Bernoulli equation is applied to determine the wind drag forces on people with different frontal areas. The well-known Du Bois body surface area formula is utilized as a function of the person's height. Also, for human beings, an approximate experimental drag coefficient of unity is used.

Chapter 5 investigates a limiting case for the Navier–Stokes equations, namely creeping flow past a sphere for different viscosity fluids. During the nineteenth century, Stokes found an exact solution for the Navier–Stokes partial differential equations for a steady, incompressible, and creeping flow past a sphere. Here I expand his work to the sensitivities of the dependent variables, such as the sphere's fall velocity and its diameter, to the governing independent variables.

In Chapter 6 the Venturi meters that are used in pipes as flow meters for incompressible fluids are analyzed. In this analysis, a steady, non-viscous and irrotational flow is assumed and again the Bernoulli principle is utilized, along with the conservation of volume flow rate. Venturi meters have been used for water and for waste water volume flow rate measurements for centuries. These gages use a converging and diverging nozzle connected in-line with a pipe. For measurement of the pressure drop in the converging nozzle, the ends of a U-tube partly filled with a measurement fluid of known density higher than that of water are attached to the upstream of the converging nozzle and to the throat area of the nozzle. There is always a correction factor, called the coefficient of discharge, between the theory and the real flow rate through the Venturi meter. The coefficient of discharge depends on the size, shape, and friction encountered in the Venturi meter.

Chapter 7 analyzes the surface shape of a fluid in a rotating cylindrical tank. In this analysis, the surface tension at the fluid's free surface and the viscous forces between the fluid and the walls of the tank are neglected. Only two forces acting on a fluid's surface particle are considered, namely the gravity force which draws the particle in the downward direction and the centrifugal force which draws the particle away from the center of rotation. Also, the spillover rotational speed is determined.

Chapter 8 investigates a pin floating on the surface of a liquid due to surface tension. Cohesive forces among the liquid molecules close to the surface of a liquid cause the surface tension phenomenon. The surface molecules of a liquid do not have similar molecules above them. As a result, these surface molecules exert greater cohesive forces on the same molecules below the surface, and those next to them on the surface. These excessive cohesive forces of the surface molecules have a tendency to contract to form a membrane-like surface and minimize their excess surface energy. The maximum pin diameters that surface tension forces will hold on different liquid surfaces were determined.

Chapter 9 tackles the steady-state behavior of small raindrops or drizzles. Mist and wind effects are neglected. Only raindrops of spherical shape are considered. For a small raindrop with diameter less than 2 mm, the gravity force balances the drag force, while neglecting the atmospheric air's buoyancy force as the raindrops fall to the Earth's surface. Experimental data is used for the drag coefficient of spherical liquid particles falling in air.

In Chapter 10 I investigate one of the most important performance parameters for different aircraft, namely their range, using Brequet's range formula. In this analysis I do not consider an aircraft's takeoff, climb, descent, and landing conditions. For aircraft with turbofan jet engines, I detail Airbus A380 and Boeing 737-800 cruising conditions. For a propeller-driven aircraft, I analyze the flight of the *Spirit of St. Louis*.

Chapter 11 treats the design of a water clock. To measure time, quite a variety of water clocks have been designed and used by humans for more than 6000 years. In this problem I analyze two water clock designs that have water flowing out of a drain hole at the bottom center of a vessel. In the first case a circular vessel's radius varies linearly with respect to time. In the second case the vessel radius is constant (i.e. a cylindrical vessel).

Water's potential energy stored behind a dam in a reservoir has been used very effectively for many decades for a spin water turbine that activates a generator to produce electricity. In Chapter 12 I apply the first law of thermodynamics for open systems, mostly identified as the modified Bernoulli equation, to sensitivities of design parameters for a water turbine system. The hydraulic diameter concept is utilized for non-circular internal flows. Friction factors are obtained from a Moody diagram, including the water tunnel's wall surface roughness.

Centrifugal acceleration forces have been used very effectively to separate solid particles from fluids or to separate different density fluids. In Chapter 13 I investigate the separation of particles in a fluid flow by centrifugal forces in a centrifuge shaped like a concentric cylinder. The fluid, along with spherical particles of different diameters, enters the centrifuge at the bottom of the centrifuge and high centrifugal forces due to a rotating inner cylinder will separate out the particles from the fluid. All particles with diameters greater than a critical diameter will adhere to the walls of the stationary outer cylinder while the fluid exits the centrifuge at the top. The exiting fluid will contain only particles with diameters smaller than the critical diameter. In this problem I use the drag forces applied to the spherical particles in a creeping flow, namely I have to make sure that the inertial forces acting on a spherical particle are much smaller than the viscous forces.

Chapter 14 deals with the analysis of a simple carburetor. The first law of thermodynamics for a perfect gas in a steady, one-dimensional and compressible flow is used to simulate the air flow in a simple carburetor. The fuel entering the throat area of the carburetor is treated as an incompressible fluid using Bernoulli's equation. The carburetor's most important dependent variable (the air–fuel ratio)'s sensitivity to independent variables is determined.

Chapter 15 analyzes properties such as the temperature, pressure, and density of an ideal gas flowing through nozzles and diffusers. Analyzing the flow of compressible fluids such as air requires use of the first and second laws of thermodynamics. The mass flow rate of an ideal gas is determined as a function of the Mach number, while assuming constant specific heats and an isentropic flow process.

Chapter 16 investigates the head loss sensitivities due to friction to the governing independent variables such as pipe diameter, fluid kinematic viscosity, and volume flow rate for steady, incompressible, and fully developed laminar flows through a straight pipe of constant cross-section. Simplified Navier–Stokes equations in cylindrical coordinates are used, by assuming that the velocity components in the tangential and radial directions are zero. The only velocity component of the fluid flowing in the pipe is in the axial direction, and it is a constant; this velocity component only varies in the radial direction. Also, the pressure gradient along the axial direction of the pipe is a constant.

Chapter 17 determines the power input requirements for a water supply line from a lake to a factory for different volume flow rates and different pipe internal diameters using the first law of thermodynamics. Friction factors are extracted from Moody diagrams for turbulent pipe flows for different Reynolds numbers and different pipe wall roughness. The head losses due to bends, valves, couplings, elbows, water entrance and exit are obtained from the manufacturer for each component. These head losses are obtained experimentally and the results depend on the Reynolds number and the component's diameter.

Chapter 18 investigates the air and water flow required to cool a printed circuit board. The components on a printed circuit board generate power, and these components have to be kept at a specified operating temperature. For the present analysis, the forced convection heat transfer coefficient is determined by making an analogy, namely the Reynolds analogy, between the wall shear stress and the rate of heat flow per unit area perpendicular to the flat plate surface.

In Chapter 19 I investigate the rate of mass convection (i.e. evaporation flux) from a flat water surface to air that is flowing over the water surface at a constant velocity. The analysis is done by assuming steady-state conditions for the convective mass transfer and constant water and air properties at 1 atm air pressure and for different water surface and air temperatures. Water vapor is assumed to behave as an ideal gas in air. Only mass transfer through the water–air interface boundary layer by convective diffusion is investigated. Also, the water surface is assumed to be a flat plate. All other types of mass diffusion exchange that can occur at the water–air interface due to radiation, condensation, conduction, and so on are neglected.

Chapter 20 investigates a natural convection heat transfer problem. I consider heating a room by natural (free) convection and by radiation using a vertical flat plate water heater which is set up perpendicular to the floor. In this analysis, all the thermophysical properties of air are determined at the film temperature. In the vertical plate water heater, the natural convection boundary layer is mixed. It starts as a laminar boundary layer at the bottom of the plate and becomes turbulent at a certain transition height. The natural convection heat transfer coefficients for this laminar plus turbulent mixed boundary layer are obtained from the experimental data available in the literature.

In Chapter 21 I consider laminar flow through porous material beds. Throughout history, flows through porous media have been studied both experimentally and theoretically. The dependent variables that have to be determined are the pressure drop through the porous material bed and the fluid volume flow rate as a function of porous material void fraction, characteristic void diameter, porous material bed length and area, and fluid viscosity. Flows through voids are assumed to behave like a fully developed laminar flow in a pipe. The sensitivities of the dependent variables with respect to the independent variables are investigated.

Chapter 22 analyzes the laminar flow of the condensate film of water vapor on the surface of a vertical plate. Condensation occurs if the water vapor is cooled below its saturation temperature. During condensation, the latent heat of condensation is released and assumed to be transferred from the water vapor–liquid condensate film interface to the cooler vertical plate surface only by conduction. The condensate film starts at zero thickness at the top of the vertical plate and starts to flow downwards under the action of gravity, while the condensate film thickness grows. In the present analysis, a laminar flow of the condensate film is assumed, namely Reynolds number less than 2000, and it is also assumed that the condensate film is flowing downwards smoothly, without any waviness.

In Chapter 23 I consider time-independent non-Newtonian fluids flowing in a pipe in a fully developed state and in a laminar flow regime. Non-Newtonian fluids are studied in rheology, which is the science of deformation and fluid flow. In Newtonian fluids like water, air, milk, and so on, the shear stress applied to a fluid element is proportional to the shear rate, namely to the local velocity gradients, where the proportionality constant is the fluid's viscosity and is a constant at a given temperature. On the other hand, for non-Newtonian fluids, the fluid's viscosity is not a constant at a given temperature and changes with the magnitude of the shear rate and in some cases with time. Most non-Newtonian fluids follow a power law relationship between the shear stress and the shear rate. The flow of mayonnaise, which obeys a power law for non-Newtonian pseudoplastic fluids, is analyzed in detail in a long pipe under fully developed laminar conditions at room temperature.

Chapter 24 considers a two-phase flow, namely carbon dioxide bubbles in gas phase, flowing upwards in a stationary glass of beer which is in liquid phase. The rising bubbles are assumed to be rigid spheres and their upward motion is governed by three forces, the upward buoyancy force, the downward gravity force (bubble weight), and the drag force. Beer bubbles that are rising to the top of the glass expand and accelerate due to an increase in the effective buoyancy force.

I would like to dedicate this book to my excellent teachers and mentors in fluid mechanics and in heat transfer at several universities and organizations. Some of the names at the top of a long list are Professor I. Flugge-Lotz, Professor W. C. Reynolds, Professor W. M. Kays, Professor A. L. London, Professor R. D. Haberstroh, Professor L. V. Baldwin, and Professor T. N Veziroglu.

1

Draining Fluid from a Tank

Many fluid mechanics problems can be solved by making several simplifying assumptions and come up with fairly accurate results. If we consider conservation of mechanical energy along a streamline in a steady flow, namely no changes are occurring with respect to time in variables such as pressure, density, and flow velocity along a streamline; frictionless flow, namely there are no shear or viscous forces acting on the fluid; an incompressible fluid, where the fluid's density is a constant and a non-rotating flow, then the governing fluid mechanics equations simplify to the following Bernoulli's equation along a fluid's streamline:

$$p + \rho \times g \times z + 0.5 \times \rho \times V^2 = \text{constant} \qquad (1.1)$$

See Ref. [13] for more details regarding Bernoulli's equation.

Bernoulli's equation conserves the sum of pressure energy p, the potential energy $\rho * g * z$, and the kinetic energy $0.5 * \rho * V^2$ along a streamline or along many streamlines, namely positions 1 and 2 as shown for a fluid like water in a large tank with a small discharge hole in Figure 1.1. Both positions 1 and 2 are at atmospheric pressure. Streamlines starting at the top of the fluid surface, position 1, converge at the small drain hole at position 2. We can apply the Bernoulli equation between positions 1 and 2 as follows:

$$[p_{atm} + \rho_{fluid} \times g \times h + 0]_1 = [p_{atm} + 0 + 0.5 \times \rho_{fluid} \times V^2]_2 \qquad (1.2)$$

$V_1 \approx 0$, since we assume a large-diameter cylindrical tank with a small discharge hole, at position 2. Also, the potential energies of the streamlines exiting the tank are zero at position 2. h is the distance from the discharge hole level to the top surface level of the water in the tank.

Then:

$$V_2 = \sqrt{2 \times g \times h} \qquad (1.3)$$

Case Studies in Fluid Mechanics with Sensitivities to Governing Variables, First Edition. M. Kemal Atesmen.
© 2019 John Wiley & Sons Ltd. This Work is a co-publication between John Wiley & Sons Ltd and ASME Press.

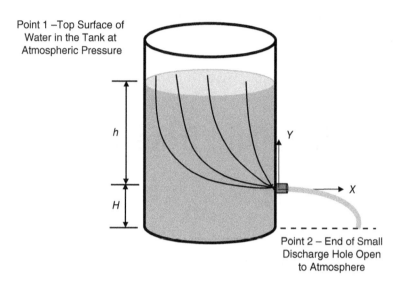

Point 1 – Top Surface of
Water in the Tank at
Atmospheric Pressure

h

Y

X

H

Point 2 – End of Small
Discharge Hole Open
to Atmosphere

Figure 1.1   Low-viscosity fluid draining from a large-diameter cylindrical tank through a small discharge hole where the total fluid height in the tank is $Z = h + H$

From Eq. (1.3), we can observe that the velocity of the steady, frictionless, incompressible (like water), and non-rotating fluid depends only on the potential energy acting on the streamlines in the tank. It is also customary to include an experimentally determined discharge velocity coefficient C_V in Eq. (1.3) in order to cover the energy losses that occur at the discharge hole due to its shape, size, material, and so on. With the inclusion of C_V, Eq. (1.3) becomes:

$$V_2 = C_V \times \sqrt{2 \times g \times h} \qquad (1.4)$$

The velocity of the fluid particles exiting the discharge hole depends on the potential energy available between positions 1 and 2. Let us now calculate the x-distance L that these fluid particles will travel from the tank discharge point before hitting the flat ground that the tank is sitting on. When a fluid particle leaves the discharge point, it has no acceleration in the x-direction and it has a negative gravitational acceleration in the y-direction, namely:

$$\frac{d^2x}{dt^2} = 0 \text{ and } \frac{d^2y}{dt^2} = -g \qquad (1.5)$$

With initial conditions in the x-direction, namely at $t = 0$, $x = 0$, and $\frac{dx}{dt} = V_2$, we obtain the x-direction position of the fluid particle, after it leaves the discharge hole, by integrating the first equation in Eq. (1.5), to be $x = V_2 \times t$. Similarly, with initial conditions in the y-direction, namely at $t = 0$, $y = 0$, and $\frac{dy}{dt} = -g \times t$, we obtain the y-direction position of the fluid particle to be $y = -0.5 \times g \times t^2$. The fluid particle hits the ground at a distance $y = -H$ or after a time $T = \sqrt{\frac{2 \times H}{g}}$. The x-distance that the fluid particle will travel is:

$$L = V_2 \times T$$

or

$$L = 2 \times C_V \times \sqrt{h \times H} \tag{1.6}$$

We can rewrite Eq. (1.6) by normalizing the x-distance L with respect to the fluid height $Z = H + h$ in the tank as follows for ease of analysis:

$$\frac{L}{Z} = 2 \times C_V \times \sqrt{\frac{H}{Z}\left(1 - \frac{H}{Z}\right)} \tag{1.7}$$

Then we can obtain the sensitivity of the x-distance L with respect to the discharge hole height H from the tank's base as follows:

$$\frac{dL}{dH} = \frac{C_V \times \left(1 - 2 \times \frac{H}{Z}\right)}{\sqrt{\frac{H}{Z} \times \left(1 - \frac{H}{Z}\right)}} \tag{1.8}$$

The normalized x-distance, $\frac{L}{Z}$, is plotted in Figure 1.2 for two different discharge velocity coefficients, namely 0.95 and 1, using Eq. (1.7). When the discharge hole is at the bottom of the tank or at the top of the fluid surface level, $L = 0$. When $\frac{H}{Z} = 0.5$, the x-distance L achieves its maximum value of $C_V * Z$.

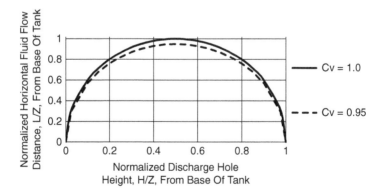

Figure 1.2 Normalized horizontal fluid flow distance, $\frac{L}{Z}$, as a function of normalized discharge hole height, $\frac{H}{Z}$.

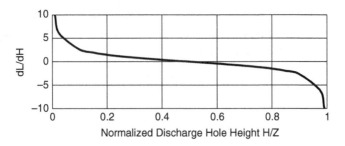

Figure 1.3 Sensitivity of horizontal fluid flow distance to discharge hole height shown as a function of normalized discharge hole height, $\frac{H}{Z}$

The sensitivity of the x-distance to discharge hole height H is plotted in Figure 1.3 using Eq. (1.8) and $C_V = 1$. The sensitivity of the x-distance to discharge hole height H is zero when the discharge hole is in the middle of the fluid height Z in the tank. Sensitivity does not change much when the normalized discharge hole height is in the range $0.4 < \frac{H}{Z} < 0.6$. The sensitivity of the x-distance to discharge hole height H changes very fast as $\frac{H}{Z}$ approaches 0 or 1.

In the above analysis, we assumed a large-diameter cylindrical tank top surface area A_1 with a small discharge hole A_2 (i.e. $V_1 \approx 0$). Now let us look at the case where $\frac{A_1}{A_2}$ is finite. The fluid mass inside the tank decreases as fluid discharges through the discharge hole. The conservation of mass for the fluid inside the tank for an incompressible fluid can be written as follows:

$$-\frac{d(A_1 \times y)}{dt} = A_2 \times V_2 \qquad (1.9)$$

Using Eq. (1.4), we can rewrite Eq. (1.9) as:

$$\frac{dy}{dt} = \frac{A_2}{A_1} \times C_V \times \sqrt{2 \times g \times y} \qquad (1.10)$$

Separating variables and integrating Eq. (1.10) from $t = 0$ at $y = h$ to $t = t^*$ at $y = h^*$ gives:

$$\int_0^{t^*} dt = -\frac{A_1}{A_2} \times \frac{1}{C_V} \times \frac{1}{\sqrt{2 \times g}} \times \int_h^{h^*} \frac{dy}{\sqrt{y}} \qquad (1.11)$$

or

$$t^* = \frac{A_1}{A_2} \times \frac{1}{C_V} \times \sqrt{\frac{2}{g}} \times (\sqrt{h} - \sqrt{h^*}) \qquad (1.12a)$$

or

$$h^*(t^*) = \left(\sqrt{h} - \frac{A_2}{A_1} \times C_V \times \sqrt{\frac{g}{2}} \times t^* \right)^2 \qquad (1.12b)$$

where $h \geq h^* \geq 0$.

The x-distance, L^*, that the fluid particle will travel before it hits the ground from the discharge hole after a certain time passes can be obtained from the combination of Eqs. (1.6) and (1.12b). L^* becomes a function of time as follows:

$$L^*(t^*) = 2 \times C_V \times \sqrt{h^*(t^*) \times H} \qquad (1.13)$$

Using Eqs. (1.12b) and (1.13) and using parameters $H = 0.5$ m, $h = 0.5$ m, $\frac{A_1}{A_2} = 1000$, and $C_V = 1$, the horizontal discharge distance, L^*, versus time, t^*, is plotted in Figure 1.4. The total time of discharge is 319.4 s. The horizontal discharge distance starts at 1 m and decreases linearly with respect to time.

Using Eq. (1.12b) and using parameters, $h = 0.5$ m, $\frac{A_1}{A_2} = 1000$, and $C_V = 1$, fluid height, h^*, versus fluid discharge time, t^* is plotted in Figure 1.5. The fluid

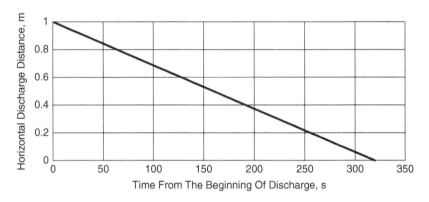

Figure 1.4 Horizontal discharge distance L^* versus time t^* from the beginning of discharge

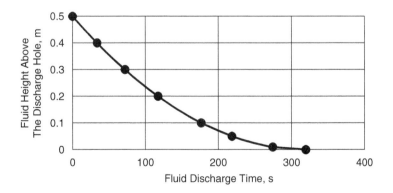

Figure 1.5 Fluid height above the discharge hole h^* versus fluid discharge time t^*

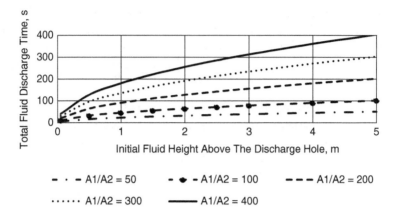

Figure 1.6   Total fluid discharge time from the tank versus initial fluid height above the discharge hole for different $\frac{A_1}{A_2}$ ratios

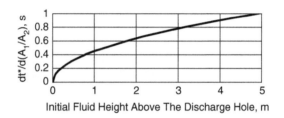

Figure 1.7   Sensitivity of total fluid discharge time above the discharge hole to $\frac{A_1}{A_2}$ versus initial fluid height above the discharge hole

discharge time decreases as the square root of the fluid height above the discharge hole.

Again using Eq. (1.12b) and using parameters $h^* = 0$ and $C_V = 1$, total fluid discharge time versus initial fluid height above the discharge hole, h, is plotted in Figure 1.6 for several different $\frac{A_1}{A_2}$ ratios. The total fluid discharge time increases as the square root of the initial fluid height above the discharge hole. As the $\frac{A_1}{A_2}$ ratio increases, the total discharge time becomes more sensitive to variations in the initial fluid height above the discharge hole.

The sensitivity of total fluid discharge time to $\frac{A_1}{A_2}$ is shown in Figure 1.7 using derivative of Eq. (1.12b) and parameters $h^* = 0$ and $C_V = 1$. This sensitivity behaves like $\sqrt{h}$ which can be deduced from Eq. (1.12b).

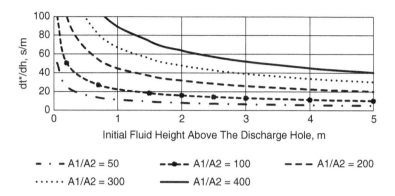

Figure 1.8 Sensitivity of total fluid discharge time to initial fluid height above the discharge hole versus initial fluid height above the discharge hole for different $\frac{A_1}{A_2}$ ratios

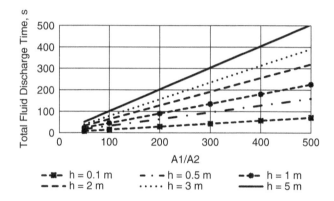

Figure 1.9 Total fluid discharge time from the tank versus $\frac{A_1}{A_2}$ for different initial fluid heights

The sensitivity of the total fluid discharge time to initial fluid height above the discharge hole is shown in Figure 1.8. This sensitivity behaves like $1/\sqrt{h}$ which can also be deduced from Eq. (1.12b) and shown in Figure 1.8 using parameters $h^* = 0$ and $C_V = 1$. This sensitivity, dt^*/dh, increases fast for small h values and approaches zero for large h values.

Using Eq. (1.12b), and using parameters $h^* = 0$ and $C_V = 1$, total fluid discharge time versus $\frac{A_1}{A_2}$ ratio for different initial fluid heights, h, above the discharge hole, is plotted in Figure 1.9. As the initial fluid height h above the discharge hole increases, the total discharge time becomes more sensitive to variations in the $\frac{A_1}{A_2}$ ratio.

2

Vertical Rise of a Weather Balloon

We are always fascinated by a rising balloon into the atmosphere. Analysis of a rising balloon can be very complicated, due to wind drag forces, decreasing air pressure and density, changing air temperature, decreasing gravity, gas used in the balloon, balloon's structural integrity, and so on. For every engineering problem we want to solve, we have to make reasonable simplifying assumptions and obtain approximate answers. In this analysis, let us assume that there are no wind drag forces on the balloon and a constant gravity all the way through the upper stratosphere, 50 km from the Earth's surface. Let us also assume that the ideal gas law applies all the way through the upper stratosphere.

First we have to determine atmospheric air density as a function of altitude, because we have to do a force balance on the balloon at the desired altitude. For the buoyancy force acting on the balloon at a desired altitude, we have to know the air density at that level. From experimental data, atmospheric temperatures as functions of altitude are given in Eqs. (2.1a), (2.1b), (2.2a), (2.2b), (2.3a), and (2.3b). Linear variations of atmospheric temperatures with altitude are also called lapse rate equations.

For the troposphere (0–11 000 m)

$$T = 15.04 - 0.00649 \times Z \, (^{\circ}C) \tag{2.1a}$$

$$T = 288.2 - 0.00649 \times Z \, (K) \tag{2.1b}$$

Case Studies in Fluid Mechanics with Sensitivities to Governing Variables, First Edition. M. Kemal Atesmen.
© 2019 John Wiley & Sons Ltd. This Work is a co-publication between John Wiley & Sons Ltd and ASME Press.

For the lower stratosphere (11 000–25 000 m)

$$T = -56.46\,(^{\circ}C) \qquad (2.2a)$$

$$T = 216.7\,(K) \qquad (2.2b)$$

For the upper stratosphere

$$T = -131.21 + 0.00299 \times Z\,(^{\circ}C) \qquad (2.3a)$$

$$T = 141.95 + 0.00299 \times Z\,(K) \qquad (2.3b)$$

Hydrostatic pressure in the atmosphere can be determined by doing a force balance to a thin column of air, as shown in Figure 2.1, where the pressure is assumed to vary only in the Z-direction.

Forces on a thin column of air can be written as:

$$P \times A = (P + dP) \times A + \rho_{air} \times g \times A \times dZ \qquad (2.4)$$

where P (Pa, or N m^{-2}) is the pressure acting on the bottom surface area of the small column of air, A (m^2) is the bottom and top surface area of the small column of air, $P + dP$ is the pressure acting on the top surface area of the small column of air, ρ_{air} is the atmospheric air density at level Z, g is the Earth's acceleration due to gravity at sea level, namely 9.8 m s^{-2}, and dZ (m) is the thickness of the small column of air. Rearranging Eq. (2.4) gives the variation of hydrostatic pressure with respect to altitude as a function of atmospheric air density:

$$\frac{dP}{dZ} = -\rho_{air} \times g \; [N\,m^{-3} \text{ or } kg\,(m^2 - s^2)^{-1}] \qquad (2.5)$$

We will assume that the atmospheric air behaves approximately as the ideal gas law, which relates the gas pressure P to its density ρ_{air}, to its temperature T (K), and to its molar mass M (g mol^{-1}), as given in Eq. (2.6). We are assuming that all ideal gas variables vary only as a function of altitude:

$$P(Z) = \rho_{air}(Z) \times T(Z) \times \frac{R}{M} \qquad (2.6)$$

Here R is the universal gas constant, which has a value of 8.206×10^{-5} (m^3 atm)(mol K)$^{-1}$, and M has a value of 28.97 g mol^{-1} for dry air.

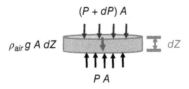

$(P + dP)\,A$

$\rho_{air}\,g\,A\,dZ$ dZ

$P\,A$

Figure 2.1   Force balance to a thin column of air in the atmosphere

Then $\left(\frac{R}{M}\right) = R_{sp\ gas}$ is the gas constant for a specific gas. For dry air, $R_{sp\ air}$ has a value of $286.9\ \text{m}^2\ (\text{s}^2\ \text{K})^{-1}$.

Combining Eqs. (2.1b), (2.5), and (2.6) for the troposphere, we get the following integral between the sea level, $Z = 0$, and altitude Z:

$$\int_{P_o}^{P} \frac{dP}{P} = \frac{g}{R_{sp\ air}} \int_0^Z \frac{dZ}{(T_o - L \times Z)} \tag{2.7}$$

where $P_o = 101\ 325$ Pa, $T_o = 288.04$ K, $L = 0.006\ 49$ K m^{-1}, and $\left(\frac{g}{R_{sp\ air} \times L}\right) = 5.263$.

The pressure variation in the troposphere as a function of altitude, Z (m), becomes:

$$P = 101\ 325 \times (1 - 2.253 \times 10^{-5} \times Z)^{5.263}\ (\text{Pa}) \tag{2.8}$$

Similarly, combining Eqs. (2.2b), (2.5), and (2.6) for the lower stratosphere, we get the following integral between the altitude $Z = 11\ 000$ m and the altitude Z:

$$\int_{P_{11\ 000}}^{P} \frac{dP}{P} = \frac{g}{R_{sp\ air} \times T_{11\ 000}} \int_{11\ 000}^{Z} dZ \tag{2.9}$$

The pressure variation in the lower stratosphere as a function of altitude, Z (m), becomes:

$$P = 22\ 630 \times e^{(1.733 - 0.000158 \times Z)}\ (\text{Pa}) \tag{2.10}$$

Similarly, combining Eqs. (2.3b), (2.5), and (2.6) for the upper stratosphere, we get the following integral between the altitude $Z = 25\ 000$ m and altitude Z up to $50\ 000$ m:

$$\int_{P_{25\ 000}}^{P} \frac{dP}{P} = \frac{g}{R_{sp\ air}} \int_{25\ 000}^{Z} \frac{dZ}{(T_{25\ 000} + L_{us} \times Z)} \tag{2.11}$$

where $L_{us} = 0.00299$ K m^{-1}

or

$$P = 2493 \times (0.655 + 1.38 \times 10^{-5} \times Z)^{-11.424}\ (\text{Pa}) \tag{2.12}$$

We now have three temperature variations, Eqs. (2.1a)/(2.1b), (2.2a)/(2.2b), and (2.3a)/(2.3b) and three pressure variations, Eqs. (2.8), (2.10), and (2.12), as functions of altitude Z. Now we can use the ideal gas equation, Eq. (2.6), to determine the air density as a function of altitude up to 50 000 m. Atmospheric pressure, temperature, and density versus altitude are plotted in Figures 2.2–2.4, respectively.

Atmospheric pressure decreases with altitude as a power function in the troposphere, exponentially in the lower stratosphere, and again as a power function in the upper stratosphere. Atmospheric pressure reaches 31% of its sea-level value of 101 325 Pa (1 atm) on top of Mount Everest, at 8848 m.

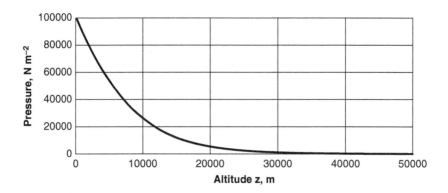

Figure 2.2   Atmospheric air pressure versus altitude

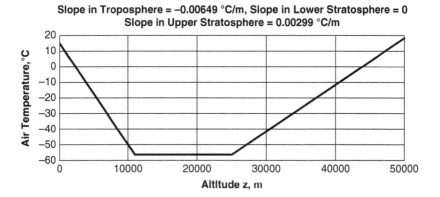

Figure 2.3   Atmospheric air temperature versus altitude

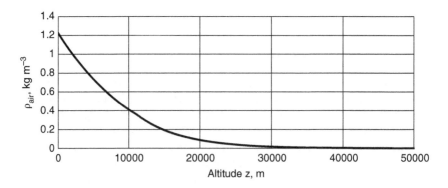

Figure 2.4   Atmospheric air density versus altitude

Atmospheric temperature decreases with altitude linearly in the troposphere, is a constant in the lower stratosphere, and increases linearly in the upper stratosphere. Air density at sea level starts at $1.226\,\mathrm{kg\,m^{-3}}$ and drops to $0.475\,\mathrm{kg\,m^{-3}}$ on top of Mount Everest, namely to 39% of sea-level value. Let us also analyze the sensitivities of atmospheric air pressure to altitude and atmospheric air density to altitude. These sensitivities are presented in Figures 2.5 and 2.6.

Both pressure and density sensitivities to altitude are very prominent at lower altitudes, but they approach zero in the upper stratosphere as we get closer to outer space.

Since we know the variation of atmospheric air density as a function of altitude, we can now try to determine the size of a helium-filled balloon that is designed to float at a given altitude. We will simplify our approach to this problem by neglecting the effects of wind, humidity, clouds, thermals, reduction in gravity, Coriolis forces, and so on. Also, in order not to deal with the balloon's volume changes and its structural integrity during its rise, we have to make two more assumptions. We will assume that the temperature of the helium gas inside the

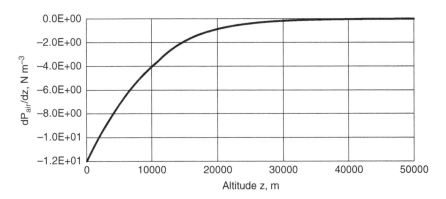

Figure 2.5 Sensitivity of atmospheric air pressure to altitude

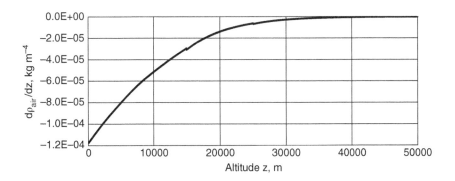

Figure 2.6 Sensitivity of atmospheric air density to altitude

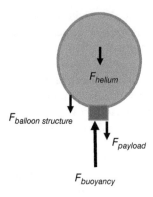

Figure 2.7   Forces acting on a floating balloon

balloon is approximately the same as the temperature of the air outside. We will also assume that the helium gas pressure is adjusted as the balloon rises, to equal the outside atmospheric pressure. A simplified force balance for a floating balloon is shown in Figure 2.7.

The forces shown in Figure 2.7 on a floating balloon can be written as:

$$F_{buoyancy} = F_{payload} + F_{helium} + F_{balloon\ structure} \tag{2.13}$$

Equation (2.13) can be expanded into:

$$\rho_{air} \times \left(\frac{4}{3} \times \pi \times R^3 \right) \times g = m_{payload} \times g + \rho_{helium} \times \left(\frac{4}{3} \times \pi \times R^3 \right) \times g$$
$$+ \rho_{mylar} \times (4 \times \pi \times R^2 \times t) \times g \tag{2.14}$$

where R (m) is the radius of the balloon, $m_{payload}$ (kg) is the payload mass, ρ_{mylar} is the density of the balloon's skin material [i.e. mylar (kg m^{-3})], and t is the thickness of the balloon's skin material. We assumed that the temperature and pressure of the helium gas inside the balloon are approximately the same as the temperature and pressure of the air outside. Using the ideal gas law, Eq. (2.6), we can relate the density of helium to the density of atmospheric air at a given altitude as:

$$\rho_{helium} = \rho_{air} \times \left(\frac{R_{sp\ air}}{R_{sp\ helium}} \right) \tag{2.15}$$

where $R_{sp\ air}$ has a value of 2077 m^2 (s^2 K)$^{-1}$.

Equation (2.14) can now be combined and written in a cubic form for the radius of the balloon, as shown:

$$R^3 - A \times R^2 - B = 0 \tag{2.16}$$

where $A = \frac{3 \times \rho_{mylar} \times t}{0.862 \times \rho_{air}}$ and $B = \frac{3 \times m_{payload}}{4 \times \pi \times 0.862 \times \rho_{air}}$.

For a given balloon skin material, skin thickness, payload, and a desired altitude of floatation where ρ_{air} is known, parameters A and B are constants. Therefore, we can find the root of Eq. (2.16) for the radius of the balloon for that desired altitude. For example, if we want to put a balloon with 20 kg of payload up to an altitude of 15 000 m, and if the balloon skin is made out of 2 mil (5.08×10^{-5} m)-thick mylar with a density of 0.194 kg m^{-3}, we can find the radius of the balloon by finding the appropriate root of Eq. (2.16). In this case, $A = 1.278$ m and $= 28.59$ m^3. For this special case the radius of the balloon is 3.55 m, as shown by the dot in Figure 2.8. It has a diameter of 7.1 m and a volume of 187 m^3.

We can now solve Eq. (2.16) for different altitudes for the same payload and balloon mylar material as above. The required balloon diameter as a function of altitude is shown in Figure 2.9. The required balloon diameter does not increase much in the troposphere, but starts to increase in the lower stratosphere. The balloon's diameter increase becomes very substantial in the upper stratosphere.

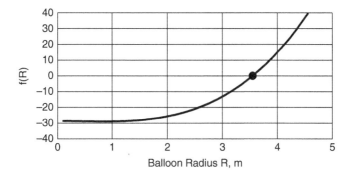

Figure 2.8 Root of Eq. (2.16) for the above special case where $f(R) = R^3 - A \times R^2 - B$

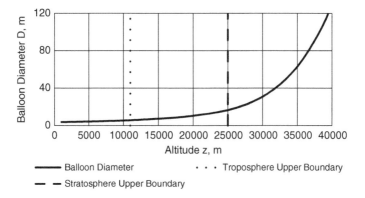

Figure 2.9 Balloon diameter as a function of altitude for a payload of 20 kg and a mylar skin material

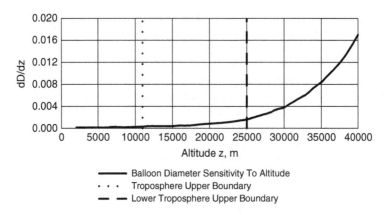

Figure 2.10   Balloon diameter sensitivity to altitude for a payload of 20 kg and a mylar skin material

Let us now analyze the sensitivity of the balloon diameter to altitude, which can be obtained from the slope of the curve in Eq. (2.9). This sensitivity is presented in Figure 2.10.

The balloon diameter sensitivity to altitude is almost zero in the troposphere. The sensitivity starts to increase in the lower stratosphere and becomes very prominent in the upper stratosphere.

3

Stability of a Floating Cone in Water

In this chapter we will analyze the stability of a right circular cone-shaped object floating in water. The untipped position of a floating cone is shown in Figure 3.1. H is the height of the cone from its apex to its base, R is the cone's base radius, h is the submerged height of the cone from its apex to the water's surface, and r is the right circular cone's radius at the water's surface. Since the center of buoyancy, B, is below the center of gravity, G, some portion of the cone is above the surface of the water. Excessive tipping of the cone under these conditions can create unstable conditions for the cone and tip it over. However, if there is sufficient restoring torque from the shifting buoyancy forces, then the cone will not tip and will restore to its original position. The vertical distance G to M is called the metacentric height. Large metacentric heights mean more restoring torque and therefore more stability. See Ref. [16], chapter 3 for a detailed analysis of stability.

Let us first find the relationship between submerged height and cone height by equating the buoyancy forces to the gravitational forces. We will neglect the buoyancy forces due to air, since there is a 1000-to-1 density ratio between water and air. Also, we will assume a constant gravity at sea level. Archimedes' principle provides us with the following force balance between buoyancy forces and gravitational forces on a body:

$$\rho_{water} \times V_{submerged} \times g = \rho_{cone} \times V_{cone} \times g \qquad (3.1)$$

From calculus courses, we can determine the volume of a right circular cone to be $V_{cone} = \frac{1}{3}\pi H R^2$ and the volume of the submerged portion of the right circular

Case Studies in Fluid Mechanics with Sensitivities to Governing Variables, First Edition. M. Kemal Atesmen.
© 2019 John Wiley & Sons Ltd. This Work is a co-publication between John Wiley & Sons Ltd and ASME Press.

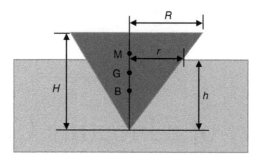

Figure 3.1   Untipped position of a floating cone in water

cone to be $V_{submerged} = \frac{1}{3}\pi h r^2$. In Figure 3.1 we can deduce that the submerged right circular cone shape is symmetrical to the total right circular cone shape, namely $\frac{h}{H} = \frac{r}{R}$. Then, the submerged height of the cone can be obtained from Eq. (3.1), as shown:

$$h = H \times \left(\frac{\rho_{cone}}{\rho_{water}}\right)^{1/3} = H \times SPG_{cone}^{1/3} \qquad (3.2)$$

where SPG_{cone} is the specific gravity of the cone material. For the cone to be in a stable region, the distance B to M has to be greater than the distance B to G. The distance B to M is defined as the second moment of inertia, $I_{axis\ of\ tipping}$, of the cone's waterline disk around the axis of its rotation (tipping) divided by the submerged volume of the cone. See Ref. [16], chapter 3 for a detailed derivation of this distance:

$$(B\ to\ M) = \frac{I_{axis\ of\ tipping}}{V_{submerged}} = \frac{(0.25\pi r^4)}{[(1/3)\pi r^2 h]} = 0.75\frac{r^2}{h} \qquad (3.3)$$

Next we have to determine the distances from the apex of the cone to the center of buoyancy and to the center of gravity. These distances can be determined in a similar manner by considering the summation of moments of disk elements to these centers. As an example, the "Apex to B" distance is calculated as follows:

$$(Apex\ to\ B) \times V_{submerged} = \int_0^h \pi x^2 z\, dz \quad where \quad \frac{x}{z} = \frac{r}{h} \qquad (3.4)$$

or

$$(Apex\ to\ B) = \left(\frac{3}{4}\right) \times h \text{ and similarly } (Apex\ to\ G) = \left(\frac{3}{4}\right) \times H \qquad (3.5)$$

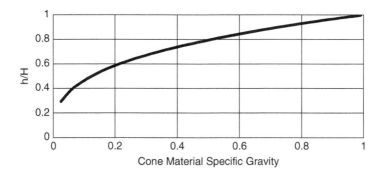

Figure 3.2 Submerged height ratio to cone height, h/H, versus cone material specific gravity

Then, the distance from the center of buoyancy to the center of gravity, B to G, is $0.75 \times (H - h)$.

For stability of the cone, combining the above equations provides us with the inequalities in Eqs. (3.6) and (3.7). In Eq. (3.6), $\frac{R}{H}$ is the independent variable and in Eq. (3.7), SPG_{cone} is the independent variable:

$$SPG_{cone} > \left[\frac{1}{1 + \left(\frac{R}{H}\right)^2} \right]^3 \tag{3.6}$$

or

$$\frac{R}{H} > \sqrt{SPG_{cone}^{-\frac{1}{3}} - 1} \tag{3.7}$$

The non-dimensional submerged height of the cone is determined from Eq. (3.2) with the specific gravity of the cone as the independent variable. The results are shown in Figure 3.2. If the right circular-shaped cone is made out of Styrofoam material with a specific gravity of 0.075, 42% of its height H will be submerged in water. If the right circular-shaped cone is made out of oak material with a specific gravity of 0.77, 92% of its height H will be submerged in water.

Next let us analyze the stability of a floating cone in water using Eq. (3.7). The region of stability is light in color and the region of instability is dark in color, as shown in Figure 3.3 as a function of the cone material's specific gravity. For low specific gravity materials (i.e. $SPG_{cone} < 0.1$), the floating cone is very unstable. For stability of such cones, the radius of the cone has to be at least as large as its height. For high specific gravity materials (i.e. $SPG_{cone} > 0.5$), the floating cone can be very stable even with a cone radius less than half its height.

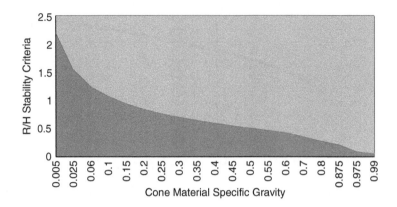

Figure 3.3 Floating cone stability criteria in water, R/H, versus cone material's specific gravity – light-colored region is stable and dark-colored region is unstable

The interface between the stable region and the unstable region is a line of neutral stability for the floating cone, namely the B to M distance is the same as the B to G distance for that particular cone.

The sensitivity of the neutral stability line, the line between the stable floating region and the unstable floating region, to changes in the cone material's specific gravity can be studied by taking the derivative of Eq. (3.7) as shown:

$$\frac{d(R/H)}{d(SPG_{cone})} = \left(-\frac{1}{6}\right) \times \frac{1}{SPG_{cone}^{7/6} \times \sqrt{1 - SPG_{cone}^{1/3}}} \qquad (3.8)$$

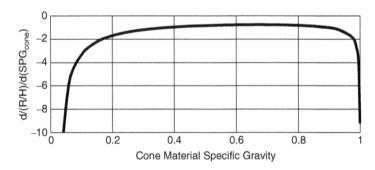

Figure 3.4 Sensitivity of floating cone stability criteria, change in R/H to change in cone material's specific gravity, versus cone material's specific gravity

As the cone material's specific gravity increases, a change in its radius-to-height ratio to be neutrally stable decreases fast with respect to a change in the cone material's low specific gravity values as shown in Figure 3.4. As the cone material's specific gravity reaches 0.2, changes in neutrally stable R/H slow down due to changes in the cone material's specific gravity and changes in neutrally stable R/H continue at a low rate until the cone material's specific gravity reaches 0.92. Between specific gravity values of 0.92 and 1.0, neutrally stable R/H values approach the limiting value of zero very fast.

4

Wind Drag Forces on People

In this chapter we are going to estimate the drag forces on a person trying to stand against the wind. His or her whole body front acts as a stagnation point for the oncoming wind. All or some of the wind's kinetic energy is converted into pressure energy. The Bernoulli equation can be applied to determine the wind drag forces on the person as follows:

$$0.5 \times V_{wind}^2 + \frac{P_{atm}}{\rho_{air}} = 0.5 \times V_{human}^2 + \frac{P_{human}}{\rho_{air}} \tag{4.1}$$

where the person is trying to stand still against the wind forces, namely $V_{human} = 0$, V_{wind} ($\mathrm{m\,s^{-1}}$) is the average wind velocity coming on to the person, P_{human} ($\mathrm{N\,m^{-2}}$) is the pressure generated by the wind forces on the person's surface area, P_{atm} is the standard atmospheric pressure at sea level (101 325 $\mathrm{N\,m^{-2}}$), and ρ_{air} at sea level is 1.2 $\mathrm{kg\,m^{-3}}$. Then, Eq. (4.1) becomes

$$(P_{human} - P_{atm}) \times A_{effective} = F_{drag} = 0.5 \times \rho_{air} \times A_{effective} \times V_{wind}^2 \tag{4.2}$$

This simple drag force model given by F_{drag} (N) and applied on an effective area of $A_{effective}$ ($\mathrm{m^2}$) in Eq. (4.2) does not apply most of the time to all surface shapes in fluid mechanics. The actual amount of force that is experienced by a body due to wind depends on its shape, flow conditions, and the body's surface roughness. Therefore, the actual amount of drag force on a body is determined experimentally and it is defined by the drag coefficient, C_{drag}, as shown in Eq. (4.3). C_{drag} values vary from 0.012 for air flow over a subsonic aircraft to 2.0 for air flow over a rectangular box:

$$F_{drag} = C_{drag} \times 0.5 \times \rho_{air} \times A_{effective} \times V_{wind}^2 \tag{4.3}$$

Case Studies in Fluid Mechanics with Sensitivities to Governing Variables, First Edition. M. Kemal Atesmen.
© 2019 John Wiley & Sons Ltd. This Work is a co-publication between John Wiley & Sons Ltd and ASME Press.

Experimental results show that the drag coefficient for a person taking on the wind in a front-facing position or in a back-facing position is about 1.0. Also, if a person is taking on the wind in a side-facing position or in a squatting position, his or her effective area reduces by about 50% of his or her frontal area and therefore the wind drag forces on that person reduce by about 50%.

Next we will relate a person's frontal area to his or her height by using the well-known Du Bois body surface area formula and the body mass index (BMI) formula given below. The Du Bois body surface formula, BSA (m^2) is

$$BSA = 0.007184 \times W^{0.425} \times H^{0.725} \tag{4.4}$$

where W (kg) is the weight of the person and H (cm) is the height of the person. A person's BMI relates their weight to their height as follows:

$$BMI = \frac{W}{\left(\frac{H}{100}\right)^2} \tag{4.5}$$

Let us consider a person with a normal weight and a BMI of 22 and let us assume that the frontal area of such a person is one-third of his or her total body surface area given by Eq. (4.4), then the effective frontal area of such a person can be determined from his or her height by combining Eqs. (4.4) and (4.5) into the following equation:

$$A_{effective} = 1.777 \times 10^{-4} \times H^{1.575} \tag{4.6}$$

Now let us analyze the wind drag forces on a person by using Eqs. (4.3) and (4.6) at different storm and hurricane wind speeds with the following assumptions:

$$C_{drag} = 1.0, \rho_{air} = 1.2 \text{ kg m}^{-2}, BMI = 22, A_{effective} = \frac{1}{3} \times BSA$$

and with the wind speed scales as given in Table 4.1.

The frontal wind drag forces on a human with BMI of 22 in a storm are shown in Figure 4.1 as a function of human height. Minimum drag forces are calculated at a wind speed of 88.4 km h^{-1}. Maximum drag forces are calculated at a wind speed of 101.4 km h^{-1}. Equation (4.5) is used to determine this person's weight as a function of his or her height. In a storm, this person can stand against the frontal wind drag forces as long as they are standing on a surface which has a static coefficient of friction greater than 0.4 between their shoe bottoms and the surface.

The frontal wind drag forces on a human with BMI of 22 in a category 4 hurricane are shown in Figure 4.2 as a function of human height. Minimum drag forces are calculated at a wind speed of 209.2 km h^{-1}. Maximum drag forces are calculated at a wind speed of 251.1 km h^{-1}. Equation (4.5) is used to determine this person's weight as a function of their height. In a category 4 hurricane this person

Table 4.1 Wind speed, V_{wind}, scales

V_{wind}	Min (km h^{-1})	Max (km h^{-1})	Min (m s^{-1})	Max (m s^{-1})	Min (mph)	Max (mph)
Storm	88.4	101.4	24.6	28.2	55	63
Violent storm	103.3	117.5	28.6	32.6	64	73
Hurricane cat 1	119.1	152.9	33.1	42.5	74	95
Hurricane cat 2	154.5	177.0	42.9	49.2	96	110
Hurricane cat 3	178.6	207.6	49.6	57.7	111	129
Hurricane cat 4	209.2	251.1	58.1	69.7	130	156
Hurricane cat 5	252.7	Above	70.2	Above	157	Above

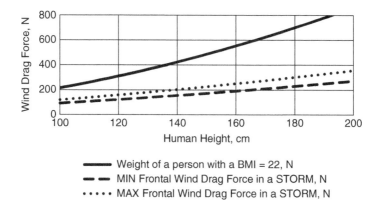

Figure 4.1 Frontal wind drag force on a human body versus human height in a storm

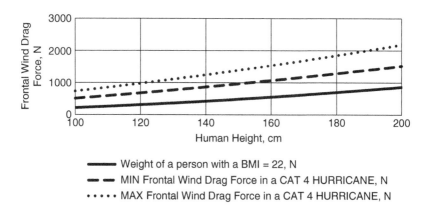

Figure 4.2 Frontal wind drag force on a human body versus human height in a category 4 hurricane

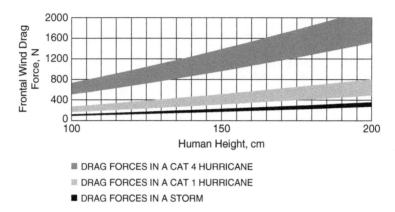

Figure 4.3 Frontal wind drag force ranges on a human body with a BMI of 22 versus human height in a storm, in a category 1 hurricane and in a category 4 hurricane

cannot stand against the frontal wind drag forces. They will be swept away in a category 4 hurricane, even if they are taking on the wind in a side-facing position or in a squatting position, where their effective area reduces by about 50%.

Let us next determine the frontal wind drag force ranges on a human body for different wind speed scales. Figure 4.3 shows these force ranges for a storm, for a category 1 hurricane, and for a category 4 hurricane. These frontal drag force ranges are calculated for a human with a BMI of 22 and an effective frontal area of one-third of his or her body surface area. Drag force ranges increase with increasing wind speed scale and with increasing effective frontal area, as expected.

Figure 4.4 shows the frontal wind drag forces versus frontal wind speed for a specific person who is 150 cm tall and has a BMI of 22. The frontal wind drag forces versus frontal wind speed for a specific person has a quadratic relationship, namely $F_{drag} = 0.285 \times V_{wind}^2$, as obtained from Eq. (4.3).

Next let us analyze the sensitivities of frontal wind drag forces to different independent variables. The change in frontal wind drag forces with respect to a change in a human's height can be determined from a combination of Eqs. (4.3) and (4.6). Again, remember that Eq. (4.6) is derived for a normal-weight person with a BMI of 22 and an effective frontal area that is one-third of his or her total body surface area. The sensitivity of the frontal drag force to human height, namely dF_{drag}/dH, is given in Eq. (4.7), and the resulting sensitivity behavior with respect to human height for different wind speed scales is shown in Figure 4.5:

$$\frac{dF_{drag}}{dH} = 1.68 \times 10^{-4} \times H^{0.575} \times V_{wind}^2 \left(\frac{N}{cm} \right) \tag{4.7}$$

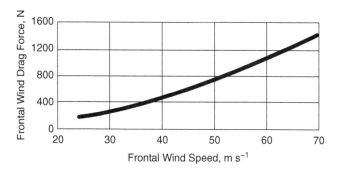

Figure 4.4 Frontal wind drag force for a 150-cm tall human with a BMI of 22 versus frontal wind speed

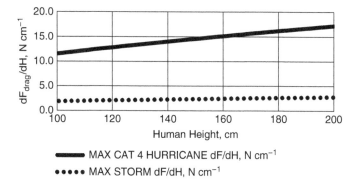

Figure 4.5 Sensitivity of frontal drag force to human height versus human height

As shown in Figure 4.5, the frontal wind drag forces are more sensitive to changes in human height at higher wind speeds.

The change in frontal wind drag force with respect to a change in a human's effective area can be determined directly from Eq. (4.3) without making any assumptions about the BMI and the effective frontal area. The sensitivity of the frontal drag force to the effective frontal area, namely $dF_{drag}/dA_{effective}$, is given in Eq. (4.8), and the resulting sensitivity behavior with respect to frontal wind speeds is shown in Figure 4.6:

$$\frac{dF_{drag}}{dA_{effective}} = 0.6 \times V_{wind}^2 \tag{4.8}$$

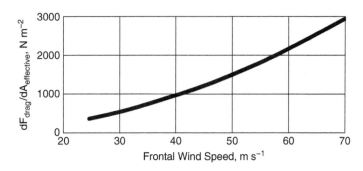

Figure 4.6   Sensitivity of a change in frontal wind drag force to a change in frontal human area at different frontal wind speeds

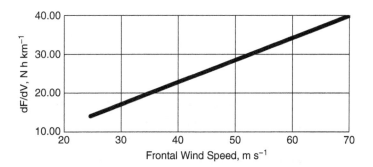

Figure 4.7   Sensitivity of a change in frontal wind drag force to a change in frontal wind speed for a 150-cm tall human with a BMI of 22 at different frontal wind speeds

A change in frontal wind drag force with respect to a change in wind speed can be determined directly from Eq. (4.3) for a specific person who is 150 cm tall and has a BMI of 22. The sensitivity of frontal drag force to wind speed, namely dF_{drag}/dV_{wind}, is given in Eq. (4.9) and the resulting linear sensitivity behavior with respect to frontal wind speeds is shown in Figure 4.7:

$$\frac{dF_{drag}}{dV_{wind}} = 0.57 \times V_{wind} \qquad (4.9)$$

5

Creeping Flow Past a Sphere

In this chapter we will investigate a limiting case for the basic differential equations, namely the Navier–Stokes equations, which are the foundation of fluid mechanics. See Ref. [15], chapter 6 for details. In this limiting case, creeping flow past a sphere, viscous forces are much greater than inertia forces (i.e. the Reynolds number is much less than 1), as shown in Eq. (5.1):

$$Re_d = \frac{Ud}{v} \ll 1 \tag{5.1}$$

where U (m s^{-1}) is the parallel stream of uniform velocity, d (m) is the diameter of the sphere, and v (m^2 s^{-1}) is the kinematic viscosity, $\frac{\mu}{\rho}$, of the fluid. Stokes found an exact solution for the Navier–Stokes partial differential equations for a steady, incompressible, and creeping flow past a sphere during the nineteenth century. See Ref. [11] for details. The following Stokes equation gives the total drag force on a sphere when the pressure and viscous forces are summed over the surface of the sphere. One-third of the total drag force on a sphere comes from pressure forces and two-thirds come from shearing viscous forces:

$$D = \sum Pressure\ Forces + \sum Viscous\ Forces \tag{5.2a}$$

$$D = (\pi \mu dU)_{pressure} + (2\pi \mu dU)_{viscous} \tag{5.2b}$$

where μ (N s m^{-2}) is the viscosity of the fluid. The results provided in Eqs. (5.2a) and (5.2b) form the basis of a falling sphere viscometer. Let us rewrite Eq. (5.2b) by defining a drag coefficient while relating the total drag force on the sphere to its frontal area and to a uniform-velocity dynamic head as follows:

$$D = C_D \times (\pi r^2) \times (0.5 \times \rho \times U^2) \tag{5.3}$$

Case Studies in Fluid Mechanics with Sensitivities to Governing Variables, First Edition. M. Kemal Atesmen.
© 2019 John Wiley & Sons Ltd. This Work is a co-publication between John Wiley & Sons Ltd and ASME Press.

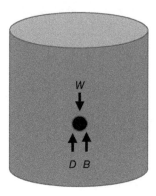

Figure 5.1 A viscometer with a falling sphere having a diameter d

where r (m) is the radius of the sphere and $C_D = \frac{24}{Re_d}$. The forces on a falling sphere viscometer are shown in Figure 5.1.

In Figure 5.1 the viscometer diameter is assumed to be large compared to the sphere's diameter, so the viscometer walls do not affect the falling sphere. W (N) is the gravity force, $\rho_{sphere} \times g \times VOL_{sphere}$, on the sphere; B (N) is the buoyancy force, $\rho_{fluid} \times g \times VOL_{sphere}$, on the sphere. The force balance for a falling sphere under steady-state conditions, namely the state where the falling sphere has reached its terminal velocity and is not accelerating, provides the following relationship:

$$W = D + B \tag{5.4}$$

Combining Eqs. (5.1), (5.2a), (5.2b), and (5.4) and rearranging gives us the following relationship for a falling sphere viscometer:

$$d^3 \ll 1.835 \times \frac{v^2}{\left(\frac{SPG_{sphere}}{SPG_{fluid}} - 1 \right)} \tag{5.5}$$

where SPG_{sphere} is the specific gravity of the sphere material and SPG_{fluid} is the specific gravity of the fluid. Let us analyze Eq. (5.5) for a small Reynolds number (i.e. $Re_d = 0.01$). Then, Eq. (5.5) takes the following form for the maximum diameter of a sphere that can be used in the viscometer, if we want to keep $Re_d \ll 1$:

$$d_{max} = 1.224 \times \frac{v^{2/3}}{\left(\frac{SPG_{sphere}}{SPG_{fluid}} - 1 \right)^{1/3}} \tag{5.6}$$

For the falling sphere viscometer to work, as can be deduced from Eq. (5.6), the specific gravity of the sphere has to be greater than the specific gravity of the fluid

Table 5.1 Maximum spherical ball diameters (mm) for creeping flow in a viscometer for $Re_d = 0.01$

	Density $(kg\,m^{-3})$	Viscosity $(N\,s\,m^{-2})$	Glass ball $SPG_{sphere} = 2.5$	Steel ball $SPG_{sphere} = 7.8$
Air @ 20°C and 1 atm	1.21	0.000018	0.058	0.040
Gasoline @ 20°C	874	0.00029	0.048	0.029
Water @ 90°C	995	0.00032	0.050	0.030
Water @ 20°C	1000	0.001	0.107	0.065
Motor oil @ 20°C	881	0.03	1.05	0.645
Motor oil @ 0°C	899	0.11	2.49	1.53
Glycerine @ 20°C	1259	1.5	13.8	7.94

with the unknown viscosity. If the fluid is water at 20°C, the specific gravity of the falling sphere has to be greater than 1. If the fluid is glycerine at 20°C, the specific gravity of the falling sphere has to be greater than 1.26. If the fluid is air at 20°C, the specific gravity of the falling sphere has to be greater than 0.001. Using Eq. (5.6), the maximum spherical ball diameters are shown in Table 5.1 for a creeping flow for $Re_d = 0.01$. Table 5.1 shows that the spherical ball diameters vary quite a lot, if the creeping flow requirements are to be met for different viscosity fluids.

Equations (5.3) and (5.4) can be combined to obtain a steady-state terminal velocity expression for the falling sphere in a fluid, as shown in Eq. (5.7):

$$U = \frac{Z}{T} = d^2 \times g \times \frac{\left(\dfrac{SPG_{sphere}}{SPG_{fluid}} - 1\right)}{(18 \times v)} \tag{5.7}$$

where Z (m) is the height that the spherical ball falls in time T (s) with the initial conditions $Z = 0$ at $T = 0$. The fall terminal velocity can be determined for different viscosity fluids using Eq. (5.7) and the creeping flow criteria (i.e. $Re_d \ll 1$). Figure 5.2 shows $\log_{10}$ of the fall terminal velocities as a function of spherical ball diameter for different viscosity fluids. In these results, the spherical ball diameters are small in order to be able to satisfy the creeping flow criterion of $Re_d \ll 1$.

Also, the spherical steel ball fall times are determined for a fall distance of 20 cm in different viscosity fluids, as shown in Figure 5.3.

Now let us analyze the effect of a falling sphere's specific gravity on the creeping flow velocity in water at 20°C. First, we have to determine the sphere diameter for a given sphere material's specific gravity from Eq. (5.6) for $Re_d = 0.01$. As the sphere material's specific gravity increases, the diameter of the sphere decreases, as shown in Figure 5.4. Also, the absolute value of the sensitivity of the sphere's diameter to the sphere material's specific gravity decreases with increasing sphere material's specific gravity, as shown in Figure 5.5.

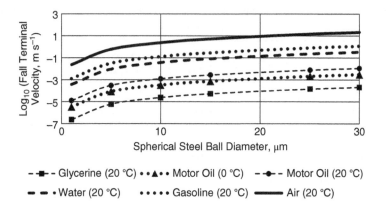

Figure 5.2 Log_{10}(fall terminal velocity) versus spherical steel ball diameter for different fluids for $Re_d \ll 1$

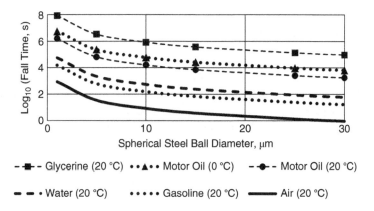

Figure 5.3 Log_{10}(fall time) in steady state for 20-cm distance versus spherical steel ball diameter for different fluids for $Re_d \ll 1$

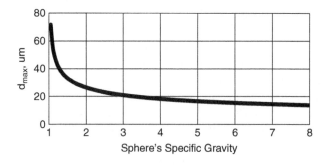

Figure 5.4 Maximum sphere diameter as a function of the sphere's specific gravity for $Re_d = 0.01$ falling in water

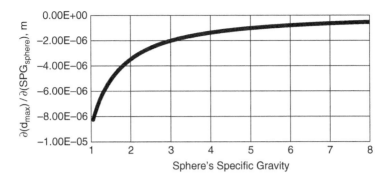

Figure 5.5 Sensitivity of sphere's diameter to sphere's specific gravity for $Re_d = 0.01$ falling in water

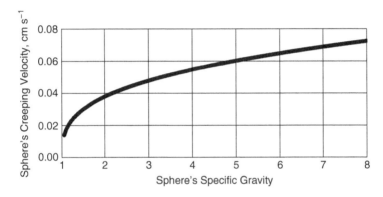

Figure 5.6 Sphere's creeping velocity falling in water as a function of sphere's specific gravity for $Re_d = 0.01$

Since we know the diameters of falling spheres for a given sphere material's specific gravity for $Re_d = 0.01$, by using Eq. (5.7) we can analyze the effects of the sphere material's specific gravity on the sphere's fall velocity. Even if the diameter of the sphere decreases with increasing sphere material's specific gravity, the fall velocity increases with increasing sphere material's specific gravity, as shown in Figure 5.6 for water at 20°C.

The sensitivity of a sphere's fall velocity to the sphere material's specific gravity decreases sharply at sphere specific gravities close to water's specific gravity at 20°C, as shown in Figure 5.7. This sensitivity diminishes considerably above sphere specific gravities of 2.

A similar analysis is performed for air at 20°C and 1 atm at $Re_d = 0.01$ and the results are shown in Figures 5.8–5.11. For a sphere falling in air, both the sphere diameter and the falling velocities are very sensitive to the sphere's specific gravity

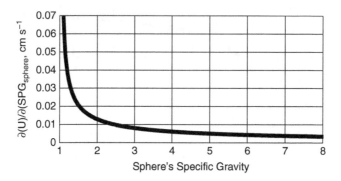

Figure 5.7   Sensitivity of sphere's fall velocity to sphere's specific gravity for $Re_d = 0.01$ falling in water

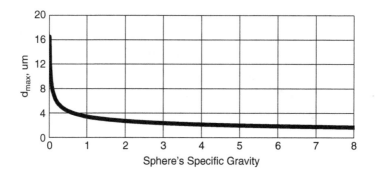

Figure 5.8   Maximum sphere diameter as a function of sphere's specific gravity for $Re_d = 0.01$ falling in air

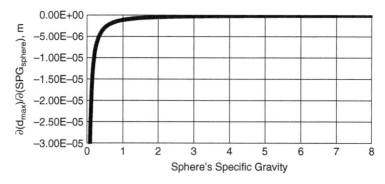

Figure 5.9   Sensitivity of sphere's diameter to sphere's specific gravity for $Re_d = 0.01$ falling in water

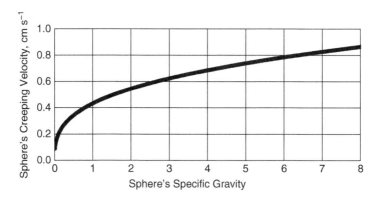

Figure 5.10 Sphere's creeping velocity falling in air as a function of sphere's specific gravity for $Re_d = 0.01$

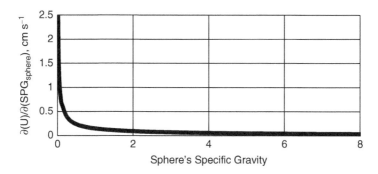

Figure 5.11 Sensitivity of sphere's fall velocity to sphere's specific gravity for $Re_d = 0.01$ falling in air

close to 0.001. Sensitivities stabilize fast and approach zero as the sphere's specific gravity increases.

The force–balance relationship given in Eq. (5.4) is applicable for a slowly falling sphere (i.e. creeping flow) in a fluid under steady-state conditions or in high-viscosity fluids where the transient effect disappears very fast after the sphere is put into the high-viscosity fluid. Equation (5.4) can be rewritten to include the transient conditions, as shown in Eq. (5.8):

$$\rho_{sphere} \times VOL_{sphere} \times \frac{d^2 Z}{dT^2} = W - D - B \qquad (5.8)$$

where Z (m) is the vertical distance traveled and T (s) is the time of travel. Equation (5.8) can be rearranged to obtain the following linear second-order differential equation:

$$\frac{d^2Z}{dT^2} + \left(\frac{18\mu}{d^2 \times \rho_{sphere}}\right) \times \frac{dZ}{dT} - \left(\frac{\rho_{sphere} - \rho_{fluid}}{\rho_{sphere}}\right) \times g = 0 \qquad (5.9)$$

with initial conditions $Z = 0$ and $\frac{dZ}{dT} = 0$ at $T = 0$. Let us introduce two parameters B and C and rewrite Eq. (5.9) as follows:

$$\frac{d^2Z}{dT^2} + B \times \frac{dZ}{dT} - C = 0 \qquad (5.10)$$

where $B = \frac{18\mu}{d^2 \times \rho_{sphere}}$ (s^{-1}) and $C = \left(\frac{\rho_{sphere} - \rho_{fluid}}{\rho_{sphere}}\right) \times g$ (m s^{-2}). Equation (5.10) can be solved easily by applying the initial conditions given above. The following fall

Table 5.2 B values for different viscosity fluids

	$B = \frac{18\mu}{d^2 \times \rho_{sphere}}$ (s^{-1})
Air @ 20°C and 1 atm	81
Gasoline @ 20°C	1 305
Water @ 90°C	1 440
Water @ 20°C	4 500
Motor oil @ 20°C	135 000
Motor oil @ 0°C	495 000
Glycerine @ 20°C	6 750 000

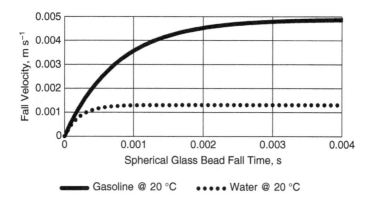

Figure 5.12   Fall velocity versus time for a 50-μm diameter spherical glass bead

velocity and vertical distance equations are obtained:

$$U = \left(\frac{C}{B}\right) \times (1 - e^{-B \times T}) \tag{5.11}$$

$$Z = \left(\frac{C}{B^2}\right) \times (e^{-B \times T} - 1) + \left(\frac{C}{B}\right) \times T \tag{5.12}$$

In Eq. (5.11), $\left(\frac{C}{B}\right)$ is the terminal velocity of the sphere under steady-state conditions. The transient term in both equations disappears fast in high-viscosity fluids, namely for large values of B. Several B values are given in Table 5.2 for different fluids with $d = 40\ \mu m$ and $\rho_{sphere} = 2500\ kg\ m^{-3}$.

The transient portion of the velocity in a creeping fall is presented for a spherical glass bead with diameter $d = 40\ \mu m$ and density $\rho_{sphere} = 2500\ kg\ m^{-3}$ in water at 20°C and in gasoline at 20°C in Figure 5.12. This glass bead reaches its terminal velocity of 0.0049 m s^{-1} in gasoline at 20°C in about 4 ms. The same glass bead reaches its terminal velocity of 0.0013 m s^{-1} in water at 20°C in less than 1 ms.

6

Venturi Meter

In this chapter we will investigate Venturi meters used in pipes as flow meters for incompressible fluids. A Venturi meter is a gage to measure the volume flow rate of a fluid in a pipe. Venturi meters have been used for water and waste water volume flow rate measurements for centuries. These gages use a converging and diverging nozzle connected in-line to a pipe. For measurement of the pressure drop in the converging nozzle, the ends of a U-tube that is partly filled with a measurement fluid of known density higher than that of water are attached to the upstream of the converging nozzle and to the throat area of the nozzle. The velocity of the fluid increases through the converging section of the nozzle, which relates to a corresponding decrease in the fluid pressure that is represented by the Bernoulli principle as given in Eq. (6.1) without any potential energy change between position 1 at the upstream of the converging nozzle and position 2 at the throat area of the nozzle. It is also assumed that the flow is steady, non-viscous, and irrotational:

$$\frac{P_1}{\rho_f} + \frac{V_1^2}{2} = \frac{P_2}{\rho_f} + \frac{V_2^2}{2} \qquad (6.1)$$

where P_1 ($N\,m^{-2}$) is the pressure at location 1, ρ_f ($kg\,m^{-3}$) is the density of the fluid, V_1 ($m\,s^{-1}$) is the fluid velocity, P_2 ($N\,m^{-2}$) is the pressure at location 2, and V_2 ($m\,s^{-1}$) is the fluid velocity at location 2. Since an incompressible fluid is assumed, the conservation of volume flow rate applies as shown in Eq. (6.2):

$$A_1 \times V_1 = A_2 \times V_2 \qquad (6.2)$$

Case Studies in Fluid Mechanics with Sensitivities to Governing Variables, First Edition. M. Kemal Atesmen.
© 2019 John Wiley & Sons Ltd. This Work is a co-publication between John Wiley & Sons Ltd and ASME Press.

where $A_1 = \left(\frac{\pi}{4}\right) \times D_1^2$ is the cross-sectional area of the circular pipe at location 1 with a diameter D_1 (m) and $A_2 = \left(\frac{\pi}{4}\right) \times D_2^2$ is the cross-sectional area of the circular Venturi throat at location 2 with a diameter D_2 (m). The static pressure balance for the U-tube between locations 1 and 2 is given by:

$$\frac{P_1 - P_2}{\rho_f \times g} = (SPG_U - 1) \times H + 0.5 \times (D_1 - D_2) \tag{6.3}$$

where $g = 9.81$ (m s^{-2}) is the gravitational constant, SPG_U is the specific gravity of the measurement fluid in the U-tube, and H (m) is the unbalanced height of the measurement fluid.

Equations (6.1)–(6.3) can provide us with the unknown flow rate in the pipe, namely $Q = A_1 \times V_1$ (m^3 s^{-1}). However, there is always a correction factor, called the coefficient of discharge, C_d, between the theory and the real flow rate through the Venturi meter. The coefficient of discharge depends on the size, shape, and friction encountered in a Venturi meter, and has a value of about 0.95. The flow rate relationship is shown in Eq. (6.4):

$$Q = C_d \times A_2 \times \sqrt{\frac{2 \times g}{(1 - D_2^4/D_1^4)} \times [(SPG_U - 1) \times H + 0.5 \times (D_1 - D_2)]} \tag{6.4}$$

Next we will analyze the Venturi meter performance for water flowing in a circular pipe using the above equations. Then we will determine the sensitivity of the flow rates to measurement fluids at different specific gravities. For the present water flow performance calculations, the following parameters and constants are used:

$$D_1 = 0.2 \text{ m}, \quad \rho_f = 1000 \text{ kg m}^{-3}, \quad V_1 = 1 \text{ m s}^{-1},$$
$$P_1 = 10 \text{ atm or } 1\,013\,000 \text{ N m}^{-2}, \quad g = 9.8 \text{ m s}^{-2},$$
$$C_d = 1.0, \quad \text{and} \quad SPG_U = 13.56$$

Most Venturi meters have throat diameters about 50% of the pipe's internal diameter. The present analysis is done all the way down to the throat diameter of choke flow. For a choked flow, water starts to vaporize at the throat at a pressure of 0.023 atm at 20°C. At this low pressure, the water velocity at the throat is 45 m s^{-1}, which corresponds to $D_2 = 0.03$ m. The U-tube fluid unbalanced height using mercury as the measurement fluid versus Venturi meter throat diameter is shown in Figure 6.1. Below $\frac{D_2}{D_1} < 0.08$, the unbalanced height of mercury in the U-tube increases very fast.

Figure 6.2 shows the water velocity at the Venturi meter throat versus the Venturi meter throat diameter. V_2 varies as the inverse square of the throat diameter.

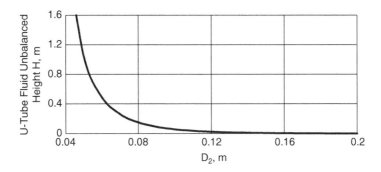

Figure 6.1   U-tube fluid unbalanced height using mercury as the measurement fluid versus Venturi meter throat diameter

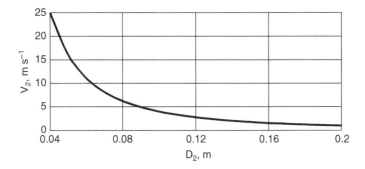

Figure 6.2   Water velocity in the Venturi meter throat versus Venturi meter throat diameter

The sensitivity of the U-tube unbalanced height to the Venturi meter throat diameter is shown in Figure 6.3. This sensitivity increases very fast with a measurement fluid of mercury below a Venturi meter throat diameter of 0.08 m.

The unbalanced measurement fluid height in the U-tube versus the Venturi meter throat diameter for measurement fluids with different specific gravities is presented in Figure 6.4. As the specific gravity value of the measurement fluid gets closer to water's specific gravity value, the unbalanced measurement fluid height increases very fast as the Venturi throat diameter decreases.

The sensitivity of the unbalanced measurement fluid height in the U-tube to the measurement fluid's specific gravity versus the Venturi meter throat diameter for different specific gravity fluids is shown in Figure 6.5. As the specific gravity value of the measurement fluid gets closer to water's specific gravity value, these sensitivities become very prominent with decreasing Venturi throat diameter. High-specific-gravity measurement fluids like mercury are less sensitive to changes in flow rates in pipes.

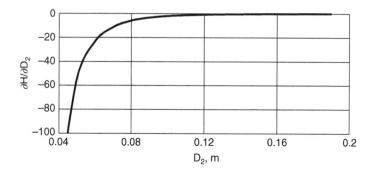

Figure 6.3 Sensitivity of U-tube unbalanced height to Venturi meter throat height for measurement fluid of mercury

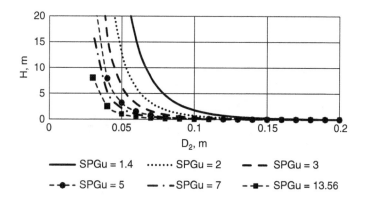

Figure 6.4 H versus D_2 for different specific gravity measurement fluids

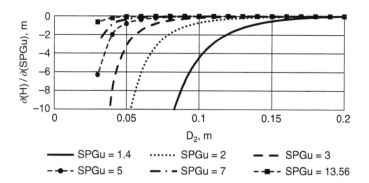

Figure 6.5 Sensitivity of U-tube measurement fluid unbalanced height to U-tube measurement fluid specific gravity for different specific gravity measurement fluids

Table 6.1   Change in U-tube measurement fluid's unbalanced height (mm) for different specific gravity measurement fluids for a Venturi meter with $D_2/D_1 = 0.5$

SPG_U	−1% Flow rate change	+1% Flow rate change
1.4	−38.07	+38.46
2.0	−15.23	+15.38
3.0	−7.61	+7.69
5.0	−3.81	+3.85
7.0	−2.54	+2.56
13.56	−1.21	+1.22

Small changes in the flow rate are analyzed for a Venturi meter with a throat diameter of 0.1 m, namely 50% of the pipe diameter. The responses of the U-tube measurement fluids' unbalanced heights to small flow rate changes are given in Table 6.1. As the specific gravity of the measurement fluid increases, the response from the U-tube measurement fluid to small flow rate changes in the pipe decreases very fast.

7

Fluid's Surface Shape in a Rotating Cylindrical Tank

In this chapter we will investigate the surface shape of a fluid in a rotating cylindrical tank, as shown in Figure 7.1. Initially, the fluid's surface is horizontal when the rotational speed of the tank is zero. Also, we will neglect the surface tension at the fluid's free surface. We will also neglect the viscous forces between the fluid and the walls of the tank. When the tank is rotating with a constant angular speed of ω (rad s^{-1}), the fluid's surface takes the form of a concave paraboloid which will be shown below when we consider only two forces acting on a fluid's surface particle, namely the gravity force which draws the particle in the negative Z-direction and the centrifugal force which draws the particle away from the center of rotation.

Let us consider an angle θ between the normal outward direction of the particle shown on the surface of the fluid in Figure 7.1 and the Z-axis. The tangent of this angle is given in Eq. (7.1) as the ratio of the centrifugal force and the gravity force acting on this particle on the fluid's surface. This angle is the same as the ratio of the change in the particle's position in the Z-direction divided by the change in the particle's position in the R-direction:

$$\tan(\theta) = \frac{dZ}{dR} = \frac{\omega^2 \times R}{g} \qquad (7.1)$$

Let us now integrate Eq. (7.1) between Z_{min}, which corresponds to the center of the cylindrical tank where there is no centrifugal force, namely at $R = 0$, and $Z_{min} + \delta$, when $R = R_{tank}$, as shown in Eq. (7.2):

$$\int_{Z_{min}}^{Z_{min}+\delta} dZ = \int_{0}^{R_{tank}} \frac{\omega^2 \times R}{g} dR \qquad (7.2)$$

Case Studies in Fluid Mechanics with Sensitivities to Governing Variables, First Edition. M. Kemal Atesmen.
© 2019 John Wiley & Sons Ltd. This Work is a co-publication between John Wiley & Sons Ltd and ASME Press.

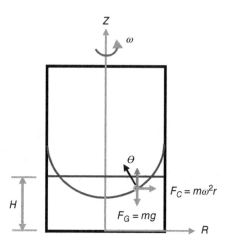

Figure 7.1   Fluid's surface shape in a rotating cylindrical tank

After integrating Eq. (7.2), we obtain the concave paraboloid Eq. (7.3) for the fluid's surface shape in a rotating cylindrical tank:

$$\delta = \frac{\omega^2 \times R^2}{2 \times g} \tag{7.3}$$

Let us now determine the distance the fluid's surface goes down at the center of the rotating cylindrical tank, namely Z_{min}, and the distance the fluid's surface goes up at the side walls of the tank to Z_{max}, where $Z_{max} = Z_{min} + \delta_{max}$ and $\delta_{max} = \frac{\omega^2 \times R_{tank}^2}{2 \times g}$. First, the volume of the paraboloid that represents the surface of the rotating fluid can be determined by integrating the elemental disk volume given in Eq. (7.4):

$$dV = \pi \times R^2 \times dZ \tag{7.4}$$

Inserting Eq. (7.1) into Eq. (7.4) gives the following equation for the elemental disk volume:

$$dV = \left(\frac{\pi \times \omega^2}{g} \right) \times R^3 \times dR \tag{7.5}$$

Integrating Eq. (7.5) from $R = 0$ to $R = R_{tank}$ gives the volume of the paraboloid:

$$\int_0^{R_{tank}} dV = \left(\frac{\pi \times \omega^2}{4 \times g} \right) \times R_{tank}^4 \tag{7.6}$$

The total volume, namely the fluid's volume plus the paraboloid volume, is given by:

$$Total\ Volume = \pi \times R_{tank}^2 \times Z_{max} = \pi \times R_{tank}^2 \times (Z_{min} + \delta_{max}) \tag{7.7}$$

The total volume minus the paraboloid volume provides the original fluid volume in the tank, given by:

$$\pi \times R^2_{tank} \times (Z_{min} + \delta_{max}) - \left(\frac{\pi \times \omega^2}{4 \times g}\right) \times R^4_{tank} = \pi \times R^2_{tank} \times H \qquad (7.8)$$

By inserting δ_{max} into Eq. (7.8), it can be shown in Eq. (7.9) that the difference between the original fluid height H and the minimum fluid height at the center of the tank Z_{min} is:

$$H - Z_{min} = \frac{\omega^2 \times R^2_{tank}}{4 \times g} = \frac{\delta_{max}}{2} \qquad (7.9)$$

Therefore, $Z_{min} = H - 0.5 \times \delta_{max}$ and $Z_{max} = H + 0.5 \times \delta_{max}$. Now the liquid surface height, $Z(R)$, in the rotating tank can be written as $Z_{min} + \delta$, as shown in Eq. (7.10):

$$Z(R) = H - \frac{\omega^2}{2 \times g} \times (0.5 \times R^2_{tank} - R^2) \qquad (7.10)$$

We are now in a position to plot the fluid's surface in a rotating cylindrical tank using Eq. (7.10). The plots shown in Figure 7.2 are for a cylindrical tank which has 1 m height and 1 m diameter. Initially, the water height in the tank is 0.5 m when the tank is stationary. The tank is rotating in the Earth's gravitational field of 9.81 m s^{-2}. The fluid surface is above the original fluid height H for radii above 0.354 m, namely $R > \frac{R_{tank}}{\sqrt{2}} = 0.354$ m, and the fluid surface is below the original fluid height H for radii below 0.354 m, independent of the centrifugal and gravity forces acting on the fluid particles.

As the rotational speed increases, the fluid surface at the walls of the tank reaches its spillover rotational speed limit. The spillover rotational speed limit can be obtained from Eq. (7.10) by setting the height of the fluid surface at $R = R_{tank}$

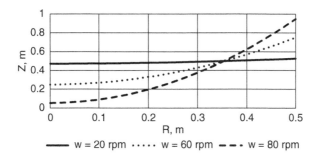

Figure 7.2 Parabolic shape of fluid surface in a cylindrical tank at three different rotational speeds for an initial water height of 0.5 m

to the tank's height, namely Z_{tank}. The spillover rotational speed limit is shown in Eq. (7.11), which is 84.6 rpm for the case presented in Figure 7.2:

$$\omega_{spillover} = \left(\frac{2}{R_{tank}}\right) \times \sqrt{g \times (Z_{tank} - H)} \qquad (7.11)$$

There is another rotational speed limit called the de-wetting limit, which occurs when the fluid surface reaches zero height at the center of the tank, namely Eq. (7.10) becomes $Z(0) = 0$. Equation (7.12) shows the de-wetting rotational speed limit, which comes out as 84.6 rpm for the case presented in Figure 7.2. For this case, the spillover and de-wetting rotational speeds are the same because the tank's height is twice the original fluid height (i.e. $Z_{tank} = 2 \times H$):

$$\omega_{de-wetting} = \left(\frac{2}{R_{tank}}\right) \times \sqrt{g \times H} \qquad (7.12)$$

The sensitivity of the fluid's surface height to the cylindrical tank radius can be obtained by differentiating Eq. (7.10), namely $\frac{\partial Z}{\partial R}$, as shown in Eq. (7.13). Plots of these sensitivities for three different rotational speeds are shown in Figure 7.3. These curves are for a cylindrical tank which has 1 m height and 1 m diameter. Initially, the water height in the tank is 0.5 m when the tank is stationary. The tank is rotating in the Earth's gravitational field of $9.81 \ m\,s^{-2}$. The sensitivities of the fluid's surface height to the cylindrical tank radius start at zero in the middle of the tank and increase linearly towards the tank's walls. As the tank's rotational speed increases, so does the sensitivity of the fluid's surface height to the cylindrical tank radius:

$$\frac{\partial Z}{\partial R} = \frac{\omega^2 \times R}{g} \qquad (7.13)$$

The sensitivity of the fluid's surface height to gravity is given in Eq. (7.14) and presented in Figure 7.4 for three different rotational speeds. The curves in Figure 7.4 are for a cylindrical tank which has 1 m height and 1 m diameter.

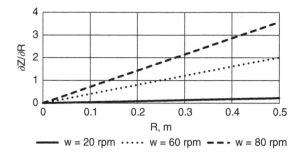

Figure 7.3   Sensitivity of fluid's surface height to cylindrical tank radius

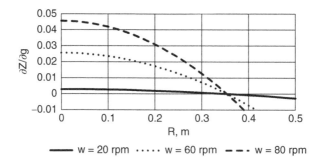

Figure 7.4 Sensitivity of fluid's surface height to gravity

Initially, the water height in the tank is 0.5 m when the tank is stationary. The tank is rotating in the Earth's gravitational field of 9.81 m s^{-2}:

$$\frac{\partial Z}{\partial g} = \frac{\omega^2}{2 \times g^2}(0.5 \times R_{tank}^2 - R^2) \tag{7.14}$$

The sensitivity of the fluid's surface height to gravity increases as the tank's rotational speed increases. The highest sensitivity is at the center of the tank. Sensitivities are positive for radii below 0.354 m, namely $R < \frac{R_{tank}}{\sqrt{2}} = 0.354$ m, and become negative for radii above 0.354 m.

Let us now investigate the spillover rotational speed limit, namely Eq. (7.11), in detail.

The spillover rotational speed limit increases as the square root of gravity, as shown in Eq. (7.11). Different space bodies have different surface gravities. Pluto, which is a dwarf planet, has a surface gravity of 0.03 m s^{-2}. Our Sun has a surface gravity of 274 m s^{-2}. The spillover rotational speed limits for different gravities are shown in Figure 7.5. Again, the curves in Figure 7.5 are for a cylindrical tank which has 1 m height and 1 m diameter. Initially, the water height in the tank is 0.5 m when the tank is stationary.

The sensitivity of the spillover rotational speed limit to gravity is obtained from Eq. (7.11) and given in Eq. (7.15). The sensitivity of the spillover rotational speed limit to gravity is shown in Figure 7.6 for the standard tank case used in this chapter. The sensitivity of the spillover rotational speed limit to gravity starts very high at low gravities and decreases towards zero as gravity increases:

$$\frac{\partial \omega}{\partial g} = \left(\frac{1}{R_{tank}}\right) \times \sqrt{\frac{Z_{tank} - H}{g}} \tag{7.15}$$

The spillover rotational speed limit versus gravity for different diameter circular tanks with a tank height of 1 m and an initial fluid height of 0.5 m is shown in

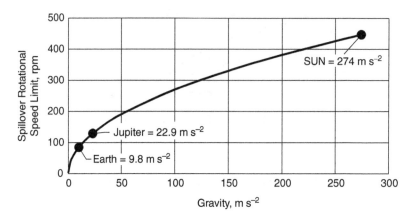

Figure 7.5   Spillover rotational speed limit versus gravity for a circular tank

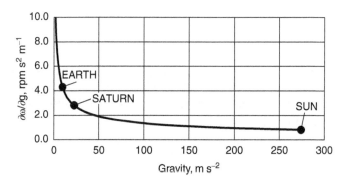

Figure 7.6   Sensitivity of spillover rotational speed limit versus gravity

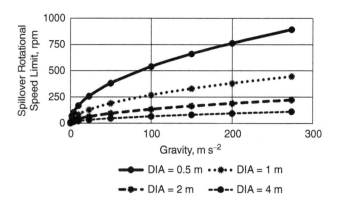

Figure 7.7   Spillover rotational speed limit versus gravity for different diameter circular tanks

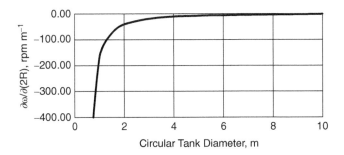

Figure 7.8   Sensitivity of spillover rotational speed limit to circular tank diameter with a tank height of 1 m and an initial fluid height of 0.5 m when the tank is stationary

Figure 7.7. As the tank's diameter increases, the spillover rotational speed limit decreases as $\frac{1}{R_{tank}}$.

The sensitivity of the spillover rotational speed limit to circular tank diameter is shown in Figure 7.8. The absolute value of this sensitivity decreases very fast for the present case. The absolute value of this sensitivity is almost zero for tank diameters above 6 m.

8

Pin Floating on Surface of a Liquid

In this chapter we will investigate a pin floating on the surface of a liquid due to surface tension. Cohesive forces among liquid molecules close to the surface of a liquid cause the surface tension phenomenon. The surface molecules of a liquid do not have similar molecules above them. As a result, these surface molecules exert greater cohesive forces on the same molecules below the surface and on those next to them on the surface. These excessive cohesive forces of the surface molecules have a tendency to contract to form a membrane-like surface and minimize their excess surface energy. Let us formulate the forces acting on a round pin with length L (m), diameter D (m), and density ρ_{pin} (kg m^{-3}) floating on the surface of a liquid, as shown in Figure 8.1.

The force balance in the vertical direction gives the following:

$$-W + F_{st1} \times \cos\theta + F_{st2} \times \cos\theta = 0 \qquad (8.1)$$

where F_{st1} and F_{st2} are the surface tension forces between the liquid's surface molecules supporting the pin on two sides and θ is the surface tension force angle between the tangent at the last liquid contact point to the pin's surface and the vertical direction. We can construct this surface tension force direction by the line that is perpendicular to the line that goes from the last liquid contact point on the pin's surface to the center point of the pin. Surface tension forces are defined for different liquids as Υ (N m^{-1}), namely force per unit length of the pin's contact surface. The weight of the pin is given by $W = \frac{\pi \times D^2}{4} \times L \times \rho_{pin} \times g$ (N), where g is Earth's gravity at sea level with a value of 9.81 m s^{-2}. The surface tension values

Case Studies in Fluid Mechanics with Sensitivities to Governing Variables, First Edition. M. Kemal Atesmen.
© 2019 John Wiley & Sons Ltd. This Work is a co-publication between John Wiley & Sons Ltd and ASME Press.

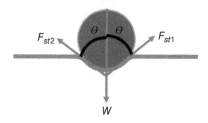

Figure 8.1   Forces acting on a round pin floating on the surface of a liquid

for liquids are obtained experimentally using different methods. For example, Υ for benzene at 20°C is $0.029\,\mathrm{N\,m^{-1}}$, for water at 20°C is $0.073\,\mathrm{N\,m^{-1}}$, for molten glass at 1200°C is about 0.3, and for mercury at 20°C is $0.486\,\mathrm{N\,m^{-1}}$. Equation (8.1) can be rewritten to find the diameter of the pin that can be supported by a liquid's surface tension forces, as shown in Eq. (8.2):

$$D = \sqrt{\frac{8 \times \Upsilon}{\pi \times \rho_{pin} \times g}} \times \cos \theta \qquad (8.2)$$

The diameter of the pin that is supported by surface tension forces is independent of the pin's length, since the surface tension force is defined per unit length of the pin. Using Eq. (8.2), the pin diameter versus surface tension force angle for three different pin materials, namely wood with a density of $700\,\mathrm{kg\,m^{-3}}$, aluminum with a density of $2600\,\mathrm{kg\,m^{-3}}$, and steel with a density of $7500\,\mathrm{kg\,m^{-3}}$, and for three different liquid surfaces, namely benzene at 20°C, water at 20°C, and mercury at 20°C, are shown in Figures 8.2–8.4.

The maximum pin diameter occurs when $\theta = 0°$ or when the liquid surface covers 180° of the pin's cross-section at its lower half. When $\theta = 90°$ the surface tension does not exert any force on the pin's surface. The sensitivity of the pin diameter to the surface tension force angle is given in Eq. (8.3). Sensitivity curves

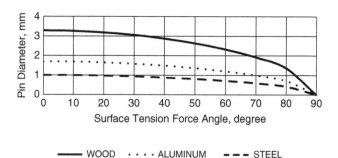

Figure 8.2   Pin diameter versus surface tension force angle floating on benzene at 20°C, $\Upsilon = 0.029\,\mathrm{N\,m^{-1}}$

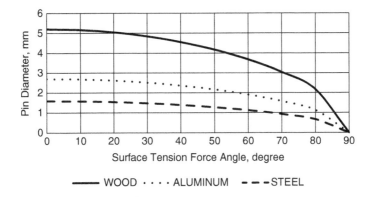

Figure 8.3   Pin diameter versus surface tension force angle floating on water at 20°C, $\Upsilon = 0.073\,\mathrm{N\,m^{-1}}$

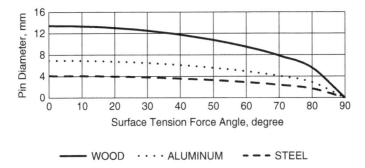

Figure 8.4   Pin diameter versus surface tension force angle floating on mercury at 20°C, $\Upsilon = 0.486\,\mathrm{N\,m^{-1}}$

for pins floating on mercury are shown in Figure 8.5. The sensitivities of pin diameters to surface tension force angle around 0° are very small, since lots of surface molecules cover almost half the pin's surface. As the pin diameter gets smaller towards zero and the surface tension force angle approaches 90°, the absolute value of the sensitivity of the pin diameter to the surface tension force angle increases very rapidly:

$$\frac{\partial D}{\partial \theta} = \frac{-\sin\theta}{\sqrt{\cos\theta}} \times \sqrt{\frac{8 \times \Upsilon}{\pi \times \rho_{pin} \times g}} \tag{8.3}$$

The maximum pin diameters that can be held by surface tension forces on different liquid surfaces are shown in Figure 8.6 for three different pin materials, namely wood, aluminum, and steel. The surface tension force angle, θ, is at zero degrees for these calculations. For example, the maximum diameter of a wooden

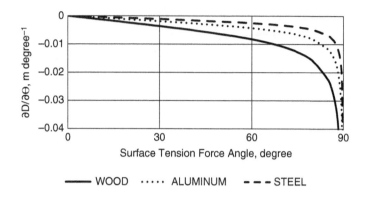

Figure 8.5 Sensitivity of pin diameter to surface tension force angle on mercury

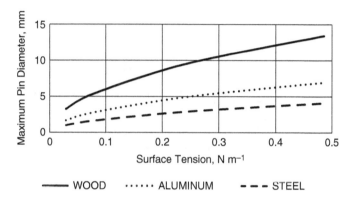

Figure 8.6 Maximum pin diameters that surface tension forces will hold on different liquid surfaces

pin with a density of $700\,\mathrm{kg\,m^{-3}}$ on a mercury surface is 13.4 mm; for a steel pin with a density of $7500\,\mathrm{kg\,m^{-3}}$ the maximum pin diameter on a mercury surface goes down to 4.1 mm.

The sensitivities of the maximum pin diameter to the liquid's surface tension can be obtained from Eq. (8.2). By setting θ equal to zero and then differentiating Eq. (8.2) with respect to Υ, we get Eq. (8.4). Figure 8.7 shows these sensitivities for three different pin materials, namely wood, aluminum, and steel. The sensitivities of changes in maximum pin diameters to changes in surface tension are high at low values of surface tension. The sensitivities shown in Figure 8.7 decrease as the surface tension values go above $0.3\,\mathrm{N\,m^{-1}}$. The sensitivities shown in Figure 8.7 also decrease considerably for pins with high densities:

$$\frac{\partial D_{max}}{\partial \Upsilon} = 0.5 \times \Upsilon^{-0.5} \times \sqrt{\frac{8}{\pi \times \rho_{pin} \times g}} \tag{8.4}$$

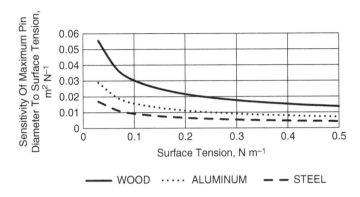

Figure 8.7   Sensitivity of maximum pin diameter to liquid's surface tension

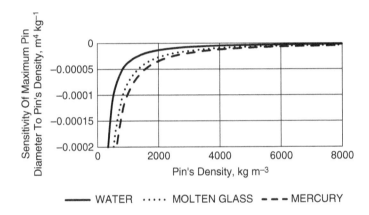

Figure 8.8   Sensitivity of maximum pin diameter to pin's density

The sensitivities of the maximum pin diameter to the pin's density can also be obtained from Eq. (8.2). By setting θ equal to zero and then differentiating Eq. (8.2) with respect to ρ_{pin}, we get Eq. (8.5). Figure 8.8 shows these sensitivities for liquids with three different surface tensions. The maximum pin diameters are very sensitive to pin density changes below pin density values of 2000 kg m^{-3}. The sensitivities shown in Figure 8.8 are almost negligible above pin density values of 4000 kg m^{-3}:

$$\frac{\partial D_{max}}{\partial \rho_{pin}} = -0.5 \times \rho_{pin}^{-1.5} \times \sqrt{\frac{8 \times \Upsilon}{\pi \times g}} \tag{8.5}$$

9

Small Raindrops

In this chapter we will investigate the steady-state behavior of small raindrops or drizzles. We will neglect the mist effect and the wind effect, and consider only raindrops that are spherical in shape. The spherical shape of a small raindrop with diameter less than 2 mm is caused by surface tension. Cohesive forces among water molecules at the surface of a small raindrop cause surface tension. These surface molecules have a tendency to contract to form a spherical shape in order to minimize their excess surface energy. As the raindrops fall through the atmosphere, they are fighting against the drag forces at their bottom surface. These drag forces tend to flatten the spherical raindrop's bottom and change the shape of the raindrop. At the same time they collide and combine with other raindrops to increase their size, while taking on irregular shapes.

For a small raindrop with diameter less than 2 mm, the gravity force balances the drag force while neglecting the atmospheric air's buoyancy force as it falls down to the Earth's surface. The force balance for a small raindrop under steady-state conditions is given as:

$$m \times g - 0.5 \times C_d \times A_{raindrop} \times \rho_{air} \times V^2_{terminal} = 0 \qquad (9.1)$$

where m (kg) is the mass of the raindrop, g is the acceleration due to gravity, which is considered a constant at $9.81 \, \mathrm{m\,s^{-2}}$, C_d is the drag coefficient that atmospheric air is exerting on the falling spherical raindrop, $A_{raindrop} = \pi \times r^2$ is the effective area of the spherical raindrop, with r being its radius in meters, ρ_{air} ($\mathrm{kg\,m^{-3}}$) is the atmospheric density of air, and $V_{terminal}$ ($\mathrm{m\,s^{-1}}$) is the raindrop's terminal velocity. The terminal velocity is the highest velocity a small raindrop reaches while falling through the atmosphere.

Case Studies in Fluid Mechanics with Sensitivities to Governing Variables, First Edition. M. Kemal Atesmen.
© 2019 John Wiley & Sons Ltd. This Work is a co-publication between John Wiley & Sons Ltd and ASME Press.

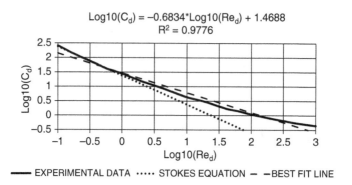

Figure 9.1   Drag coefficient for a sphere versus Reynolds number for low Reynolds numbers

The drag coefficient for spherical liquid particles falling in air has been investigated extensively by experiments. Experimental data is taken from Ref. [4]. Experimental data, best fit power equation to experimental data, and Stokes' result for creeping flow around a sphere (as treated in Chapter 5) are shown as the drag coefficient C_d versus the Reynolds number $Re_d = \frac{\bar{V} \times 2r}{\nu}$ on a $\log_{10}$–$\log_{10}$ scale in Figure 9.1.

The Stokes equation for the drag coefficient is applicable for $Re_d \ll 1$ and it is $C_d = \frac{24}{Re_d}$. The best fit power equation to experimental data, shown in Figure 9.1, is $C_d = 29.43 \times Re_d^{-0.6834}$. The coefficient of determination for this power equation's fit to the experimental data is very good (i.e. $R^2 = 0.9776$). In the present analysis, $\nu \,(\text{m}^2\,\text{s}^{-1})$ is the kinematic viscosity of atmospheric air. The best fit power equation to the experimental data can be inserted in Eq. (9.1) and the resulting expression can be rearranged to get the following terminal velocity expression for a small raindrop:

$$V_{terminal} = \left(C_1 \times r^{1.6834} \times g \times \frac{\rho_{raindrop}}{\rho_{air}} \times \nu^{-0.6834} \right)^{C_2} \tag{9.2}$$

where $C_1 = 0.1455$ and $C_2 = 0.7595$ are constants and $\rho_{raindrop}$ is the density of a small raindrop, which is assumed to be constant at $1000 \,\text{kg}\,\text{m}^{-3}$. Using Eq. (9.2), the terminal velocity for small raindrops can be determined by assuming an average atmospheric air density of $1.0 \,\text{kg}\,\text{m}^{-3}$ and an average air kinematic viscosity of $1.5 \times 10^{-5} \,\text{m}^2\,\text{s}^{-1}$. The terminal velocity versus diameter for small spherical raindrops is shown in Figure 9.2. A 0.5-mm diameter raindrop has a terminal velocity of $2 \,\text{m}\,\text{s}^{-1}$. The terminal velocity increases non-linearly as the raindrop diameter increases (i.e. $V_{terminal} \sim r^{1.28}$).

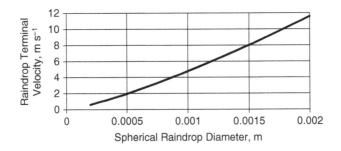

Figure 9.2 Terminal velocity versus diameter for small spherical raindrops

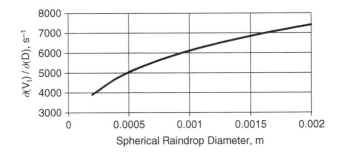

Figure 9.3 Terminal velocity sensitivity to diameter for small spherical raindrops

The sensitivity of a spherical raindrop's terminal velocity to its diameter is shown in Figure 9.3. The sensitivity is more prominent at small raindrop diameters, and decreases as the raindrop diameter increases.

Rain clouds form about 3 km above the Earth's surface in mild climates. Atmospheric air's kinematic viscosity increases from 1.46×10^{-5} m^2 s^{-1} at the Earth's surface to 1.86×10^{-5} m^2 s^{-1} at 3000 m altitude due to decreasing air density. Also, the terminal velocity of a small-diameter raindrop is higher at higher elevations. The terminal velocity sensitivity to air kinematic viscosity versus altitude for different diameter raindrops is shown in Figure 9.4. The curves in this graph are obtained by differentiating $V_{terminal}$ with respect to v in Eq. (9.2). The absolute value of the terminal velocity sensitivity to air kinematic viscosity decreases with altitude. This sensitivity becomes more prominent as the raindrop diameter increases.

The atmospheric air density decreases from 1.225 kg m^{-3} at the Earth's surface to 0.856 kg m^{-3} at 3000 m altitude in mild climates. The terminal velocity sensitivity to air density versus altitude for different diameter raindrops is shown

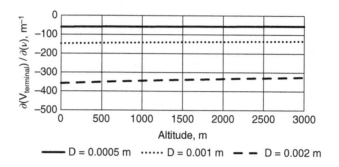

Figure 9.4 Terminal velocity sensitivity to air kinematic viscosity versus altitude for different diameter raindrops

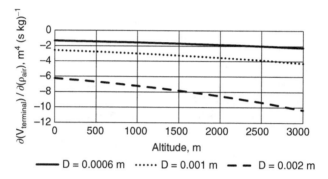

Figure 9.5 Terminal velocity sensitivity to air density versus altitude for different diameter raindrops [m^4 (kg s)$^{-1}$]

in Figure 9.5. The curves in this graph are obtained by differentiating $V_{terminal}$ with respect to ρ_{air} in Eq. (9.2) while holding the raindrop density constant at $1000\,kg\,m^{-3}$. The absolute value of the terminal velocity sensitivity to air density increases with altitude. This sensitivity becomes more prominent as the raindrop diameter increases.

10

Range of an Aircraft

In this chapter we will investigate one of the most important performance parameters for an aircraft, namely its range. In the present analysis we will not consider an aircraft's takeoff, climb, descent, or landing conditions. We will only consider the cruising conditions for different aircraft. For aircraft with turbofan jet engines, we will detail Airbus A380 and Boeing 737-800 cruising conditions. For a propeller-driven aircraft, we will analyze the *Spirit of St. Louis*.

In a cruising flight at constant speed and instantaneously adjusting elevation, we will assume that an aircraft engine's thrust T is equal to the aircraft's drag D. Also in a cruising flight at constant speed and at an instantaneously adjusting elevation, we will assume that the aircraft's weight W is equal to the aircraft's lift L (see Ref [10]). All forces are in newtons. The range R (km) is defined as the area under the cruising velocity–time curve, as given in Eq. (10.1):

$$R = \int V \times dt \tag{10.1}$$

where V ($\mathrm{km\,h^{-1}}$) is the cruising velocity of the aircraft and t (h) is the cruising time. The change in an aircraft's weight with respect to time can be expressed as the fuel consumption rate, as shown in Eq. (10.2):

$$\dot{m}_{fuel} = \frac{dW}{dt} \tag{10.2}$$

Also, an aircraft's thrust can be expressed in terms of its lift-to-drag ratio, as shown in Eq. (10.3):

$$T = \frac{W}{\left(\dfrac{L}{D}\right)} \tag{10.3}$$

Case Studies in Fluid Mechanics with Sensitivities to Governing Variables, First Edition. M. Kemal Atesmen.
© 2019 John Wiley & Sons Ltd. This Work is a co-publication between John Wiley & Sons Ltd and ASME Press.

Throughout the cruising flight at constant airspeed, we will assume a constant lift-to-drag ratio, namely the aircraft's altitude is adjusted as its weight is reduced by consumed fuel. So, with the altitude adjustment, the instantaneous lift-to-drag ratio stays constant. Combining Eqs. (10.1)–(10.3) provides the following range integral from the initial overall weight of the aircraft $W_{initial}$ to its overall final weight W_{final}:

$$R = \frac{-V \times \left(\frac{L}{D}\right)}{\left(\frac{\dot{m}_{fuel}}{T}\right)} \times \int_{W_{initial}}^{W_{final}} \frac{dW}{W} \tag{10.4}$$

The minus sign in Eq. (10.4) is needed, since the range is positive and the final weight of the aircraft is less than its initial weight. Let us introduce another very important performance criterion for turbofan jets. This performance criterion is the fuel consumption rate per unit of thrust, namely $TSFC = \frac{\dot{m}_{fuel}}{T}$ [kg (h N)$^{-1}$]. A lower TSFC value for a turbofan jet engine, namely a high-efficiency engine, is always desired. Present-day turbofan jet engines have TSFCs in the range of 0.04–0.06 kg (h N)$^{-1}$.

Equation (10.4) can be integrated to obtain Brequet's range formula for an aircraft with turbofan jets:

$$R = \frac{V \times \left(\frac{L}{D}\right)}{TSFC \times g} \times LN \left(\frac{W_{initial}}{W_{final}}\right) \tag{10.5}$$

where g is the gravitational constant, at 9.81 m s^{-2}. Since $W_{final} = W_{initial} - W_{fuel}$, Eq. (10.5) can be expressed in terms of the amount of fuel capacity an aircraft can have to start with, as shown in Eq. (10.6):

$$R = \frac{V \times \left(\frac{L}{D}\right)}{TSFC \times g} \times LN \left(\frac{1}{1 - \frac{W_{fuel}}{W_{initial}}}\right) \tag{10.6}$$

Now let us calculate the range of an Airbus A380 aircraft with the following performance parameters:

$$\frac{L}{D} = 17, \quad TSFC = 0.062 \text{ kg (h N)}^{-1}, \quad V = 918 \text{ km h}^{-1},$$

$$W_{fuel} = 256\,800 \text{ kg}, \quad W_{initial} = 575\,000 \text{ kg}$$

Using Eq. (10.6), with the above performance parameters, we get a range of 15 200 km for the Airbus A380. The time-of-flight for cruising at constant speed and constant lift-to-drag ratio for the Airbus A380 is 16.56 h. This time-of-flight is also called the endurance parameter for an aircraft without take-off, climb, descent, and landing times.

Next let us calculate the range of a short-range aircraft, namely the Boeing 737-800, with the following performance parameters:

$$\frac{L}{D} = 17, \quad TSFC = 0.039 \, \text{kg} \, (\text{h N})^{-1}, \quad V = 930 \, \text{km} \, \text{h}^{-1},$$

$$W_{fuel} = 21\,076 \, \text{kg}, \quad W_{initial} = 79\,016 \, \text{kg}$$

Again using Eq. (10.6), with the above performance parameters, we get a range of 5700 km for the Boeing 737-800. The time-of-flight for cruising at constant speed and constant lift-to-drag ratio for the Boeing 737-800 is 7.1 h.

Next let us analyze the range of a popular single-engine airplane in history, namely Charles Lindbergh's *Spirit of St. Louis*, which made the first solo flight across the Atlantic Ocean from New York to Paris in 1927.

For propeller-driven aircraft, the performance criterion used is the fuel consumption rate per unit net power, namely $PSFC = \frac{\dot{m}_{fuel}}{P_{net}}$ [kg (h W)$^{-1}$]. A low PSFC means a high-efficiency engine and high-efficiency propellers on a propeller-driven aircraft. The net power that is useful in flying the propeller-powered aircraft is defined by accounting all losses that occur in the engine, in the driveshaft, in the propellers, and so on, namely $P_{net} = \eta_{overall \, losses} \times P_{engine}$, where $\eta_{overall \, losses}$ can vary from 0.85 to 0.92. The range in Eq. (10.6) can be rewritten for propeller-driven aircraft as given in Eq. (10.7), assuming an instantaneously constant lift-to-drag ratio throughout the cruising flight at constant airspeed:

$$R = \frac{\left(\frac{L}{D}\right)}{PSFC \times g} \times LN \left(\frac{1}{1 - \frac{W_{fuel}}{W_{initial}}} \right) \qquad (10.7)$$

Using Eq. (10.7), the range of the *Spirit of St. Louis* is calculated by using the following performance parameters:

$$\frac{L}{D} = 4.73, \quad PSFC = 0.00029 \, \text{kg} \, (\text{h W})^{-1}, \quad V = 168 \, \text{km} \, \text{h}^{-1},$$

$$W_{fuel} = 1230 \, \text{kg}, \quad W_{initial} = 2015 \, \text{kg}$$

Using Eq. (10.7), with the above performance parameters, we get a range of 5632 km for the *Spirit of St. Louis*. The time-of-flight for cruising at constant speed and constant lift-to-drag ratio for the *Spirit of St. Louis* is 33.5 h. For his very long and very bumpy solo flight, Charles Lindbergh must have trained himself to stay awake and alert for at least a day and a half.

Now let us investigate the sensitivities of an aircraft's range to its performance parameters. Range varies linearly with the cruising velocity and with the lift-to-drag ratio. As an aircraft's cruising velocity and its lift-to-drag ratio increase, so does its range. Range varies inversely with TSFC or PSFC. Whenever an aircraft's thrust-specific fuel consumption (power-specific consumption for propeller airplanes) decreases, its range increases.

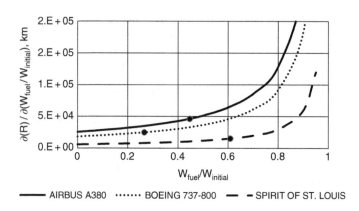

Figure 10.1 Sensitivity of range to $\frac{W_{fuel}}{W_{initial}}$ (km)

A very important performance characteristic is how much fuel an aircraft can carry to start with, namely $\frac{W_{fuel}}{W_{initial}}$. We can find the sensitivity of an aircraft's range to $\frac{W_{fuel}}{W_{initial}}$ by differentiating Eq. (10.6) for turbofan jet-powered aircraft and Eq. (10.7) for propeller-powered aircraft with respect to $\frac{W_{fuel}}{W_{initial}}$. The sensitivity of a turbofan jet-powered aircraft's range to $\frac{W_{fuel}}{W_{initial}}$ is shown in Eq. (10.8):

$$\frac{\partial R}{\partial \left(\frac{W_{fuel}}{W_{initial}} \right)} = \frac{V \times \left(\frac{L}{D} \right)}{TSFC \times g} \times \left(\frac{1}{1 - \frac{W_{fuel}}{W_{initial}}} \right) \tag{10.8}$$

The sensitivities of the ranges for the Airbus A380, Boeing 737-800, and *Spirit of St. Louis* to $\frac{W_{fuel}}{W_{initial}}$ are shown in Figure 10.1. Increases in aircraft ranges to increases in $\frac{W_{fuel}}{W_{initial}}$ are slow at low $\frac{W_{fuel}}{W_{initial}}$ values, but range increases become more prominent to increases in $\frac{W_{fuel}}{W_{initial}}$ at high $\frac{W_{fuel}}{W_{initial}}$ values.

The interesting design points for each type of aircraft are highlighted in Figure 10.1. For the Airbus A380, $\frac{W_{fuel}}{W_{initial}} = 0.45$, for the Boeing 737-800, $\frac{W_{fuel}}{W_{initial}} = 0.27$, and for the *Spirit of St. Louis*, $\frac{W_{fuel}}{W_{initial}} = 0.62$. Charles Lindbergh had to put in extra fuel tanks in order to be able to cross the Atlantic Ocean, with a high initial fuel weight-to-overall initial aircraft weight ratio.

11

Designing a Water Clock

To measure time, quite a variety of water clocks have been designed and used by humans for more than 6000 years. In this chapter we will analyze two water clock designs that have water flowing out of a drain hole at the bottom center of a vessel. In the first case, a circular vessel's radius will vary linearly with respect to time. In the second case, the vessel radius will be constant (i.e. a cylindrical vessel).

We will again consider conservation of mechanical energy along a streamline similar to the problem in Chapter 1, in a steady flow, namely no changes are occurring with respect to time in variables such as pressure, density, and flow velocity along a streamline from the top of the vessel to the drain hole at the bottom center of the vessel. We will also assume a frictionless and non-rotating flow of an incompressible fluid such as water, then the governing fluid mechanics equations simplify to the Bernoulli equation along a fluid's streamline, as discussed in Chapter 1. The velocity of the fluid particles exiting the discharge hole at the bottom center of the vessel is time dependent, and depends on the potential energy available between the positions at the top of the water level in the vessel and the drain hole at the bottom center of the vessel. We will also consider a constant discharge velocity coefficient C_V for energy losses in the drain hole. Then, the time-dependent discharge velocity $V_{discharge\ hole}(t)$ can be written as a function of the time-dependent water level h in the vessel, as shown in Eq. (11.1):

$$V_{discharge\ hole}(t) = C_V \times \sqrt{2 \times g \times h(t)} \qquad (11.1)$$

The acceleration due to gravity, g, is taken as $9.81\ \mathrm{m\,s^{-2}}$. In addition to conservation of mechanical energy, we also have to consider the conservation of mass in

Case Studies in Fluid Mechanics with Sensitivities to Governing Variables, First Edition. M. Kemal Atesmen.

the water clock vessel. Since constant density of water is assumed, conservation of mass reduces to conservation of volume for water in a circular vessel, as follows:

$$-\pi \times R^2 \times \frac{dh}{dt} = A_{drain} \times V_{discharge\ hole} \qquad (11.2)$$

where R can vary with time along with the water level h, or can be constant. We will also consider a circular drain hole area $A_{drain} = 0.25 \times \pi \times D^2$, with D being the drain hole diameter. The minus sign comes from the decreasing water level with increasing time.

In the first design case, the water level in the vessel will be assumed to vary linearly with respect to time, or $\frac{dh}{dt} = -\alpha$, where α (m s^{-1}) is a constant. If the initial condition of the water level is $h = H$ at $t = 0$, then the water level with respect to time becomes $h = H - \alpha \times t$, where H (m) is the height of the water clock vessel.

For the first design case, the water clock vessel radius, varying with respect to time, can be obtained by combining Eqs. (11.1) and (11.2) into the following Eq. (11.3):

$$R(t) = \left(\frac{A_{drain} \times C_V}{\pi \times \alpha} \right)^{0.5} \times [2 \times g \times (H - \alpha \times t)]^{0.25} \qquad (11.3)$$

We can now investigate the water clock design parameters for the first case using Eq. (11.3). First, the water level rate α has to be defined. In this example, let us assume that the water level will reduce 1 m in 6 h, namely $\alpha = 0.167$ m h^{-1}. If we want to design a 12-h clock, the height of the circular vessel has to be 2 m, namely $H = 2$ m. Next, we have to pick a drain hole size. Let us assume a cylindrical drain hole with a diameter of 2 cm. We will also assume a constant discharge velocity coefficient of 0.9, namely $C_V = 0.9$. At time equal to zero, with these parameters, the radius of the circular vessel at its top should be 3.49 m. At time equal to zero, Eq. (11.3) can be plotted for top radius versus height for different drain hole diameters, as shown in Figure 11.1.

We can see from Figure 11.1 that the top vessel radius becomes very sensitive to vessel heights below 0.5 m, namely for water clocks designed to work for less than 3 h. For the present 12-h water clock design, the water level varies linearly with time, $\alpha = 0.167$ m h^{-1}, as shown in Figure 11.2.

Also, for the 12-h water clock, the vessel radius starts at 3.49 m and reduces to zero as a function of water level, as shown in Figure 11.3. In this figure, we can see the required half shape of the 12-h water clock for the present design parameters.

The full shape of the circular 12-h water clock vessel and equally spaced water level dash marks for every 2 h on the vessel are shown in Figure 11.4.

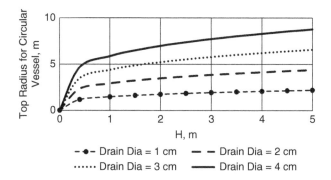

Figure 11.1 For circular vessel water clocks: top radius versus height with $\alpha = 0.167$ m h^{-1}, for different drain hole diameters and $C_V = 0.9$

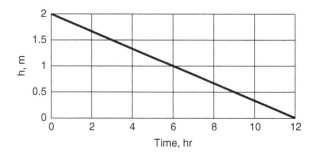

Figure 11.2 Water level in the 12-h water clock vessel versus time for $\alpha = 0.167$ m h^{-1}.

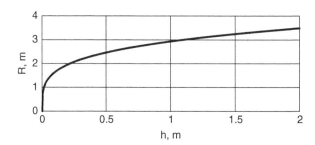

Figure 11.3 Water clock circular vessel radius versus water level for a 12-h clock with $\alpha = 0.167$ m h^{-1}

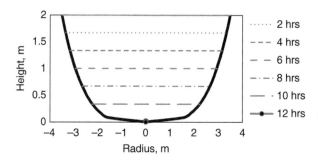

Figure 11.4 12-h water clock height versus radius for $\alpha = 0.167$ m h^{-1}, for cylindrical drain diameter 1 cm and $C_V = 0.9$

In the second water clock design case, let us consider a constant vessel radius with respect to time (i.e. a cylindrical vessel). Using Eq. (11.2) and separating water level and time variables, we get the following Eq. (11.4) in differential form:

$$\frac{dh}{\sqrt{h}} = -\left(\frac{A_{drain} \times C_V \times \sqrt{2 \times g}}{\pi \times R^2} \right) \times dt \qquad (11.4)$$

Equation (11.4) can be integrated from $h = H$ at $t = 0$ to $h = h^*$ at $t = t^*$. After rearranging the integrated equation, the water level h^* in the cylindrical vessel can be shown to be a function of time t^*, as in Eq. (11.5):

$$h^* = \left[\sqrt{H} - \left(\frac{A_{drain} \times C_V \times \sqrt{0.5 \times g}}{\pi \times R^2} \right) \times t^* \right]^2 \qquad (11.5)$$

For the present cylindrical water clock design example, we will use Eq. (11.5) with the following parameters:

Cylinder height $H = 1$ m
Cylinder radius $R = 0.5$ m
Circular drain hole diameter $= 1.4$ cm
Discharge velocity coefficient $C_V = 0.9$

This cylindrical water clock empties in 42.66 min, namely the time at $h^* = 0$. This is a water clock design for a short time duration. The water clock vessel's design and the timeline dashes are shown in Figure 11.5. The timeline dashes are not at equal intervals of time for this example, since $\frac{dh^*}{dt^*}$ is not a constant with respect to time.

We can calculate the sensitivity of the time taken to empty all the water from the vessel with respect to circular drain hole diameter. First, we have to set $h^* = 0$

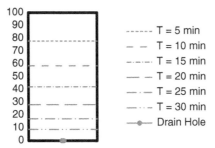

Figure 11.5 A cylindrical water clock with $R = 0.5$ m and $H = 1.0$ m with a cylindrical drain hole at the bottom center with a diameter of 1.4 cm

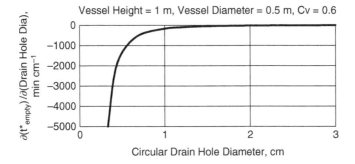

Figure 11.6 Sensitivity of time it takes to empty the cylindrical vessel with respect to circular drain hole diameter at the bottom center of the vessel

in Eq. (11.5). Then, we have to take the derivative of t^* with respect to the drain hole diameter. The sensitivity of the water emptying time to the drain hole diameter is shown in Figure 11.6. With the present design parameters, the changes in water emptying times for circular drain hole diameters less than 1.0 cm increase very fast.

We can do a similar analysis of the sensitivity as above for the velocity discharge coefficient. The results are presented in Figure 11.7. We see that more energy losses in the drain hole (i.e. smaller C_V) increases faster the time that it takes to empty the water clock vessel for the present design parameters.

Next, let us use Eq. (11.5) to observe how water level decreases with time for different vessel heights for the present constant vessel radius design. The results are shown in Figure 11.8. A cylindrical vessel with height 0.5 m has no water left in it after 30 min. A cylindrical vessel with height 1.0 m has 8.8 cm high water left in it after 30 min. A cylindrical vessel with height 1.5 m has 27.2 cm high water left in it after 30 min. The sensitivities of the water levels with respect to time,

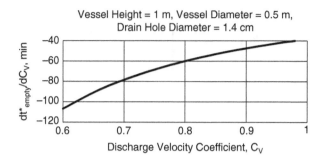

Figure 11.7 Sensitivity of time it takes to empty the cylindrical vessel with respect to discharge velocity coefficient

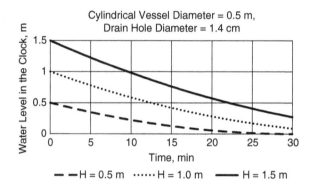

Figure 11.8 Water level versus time for different vessel heights

namely $\frac{\partial h^*}{\partial t^*}$, are shown in Figure 11.9 for three different cylindrical vessel heights. The sensitivities decrease linearly with time as the water level, potential energy, decreases in the vessel. As the vessel height increases, the changes in water level become more sensitive to changes in time.

We can also analyze different sensitivities of water levels to the present design parameters using Eq. (11.5). The sensitivities of water levels with respect to vessel heights are shown as a function of time for three different vessel heights in Figure 11.10. The water level sensitivity to vessel height decreases slowly with changing time for taller vessel heights due to larger water volumes in the vessel.

The sensitivities of water levels with respect to cylindrical vessel diameters, $\partial(h^*)/\partial(Vessel\ Diameter)$, are shown as a function of time for three different vessel diameters in Figure 11.11. The water level sensitivity to vessel diameter also decreases slowly with time for bigger vessel diameters, as expected, due to the larger water volumes in the vessel.

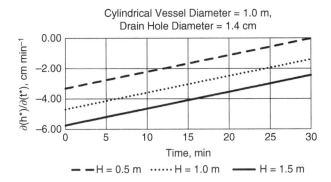

Figure 11.9 Sensitivity of water level to time after drain is opened for different vessel heights

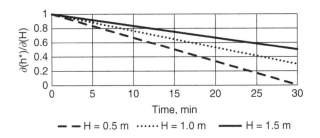

Figure 11.10   Sensitivity of water level to cylindrical vessel's height

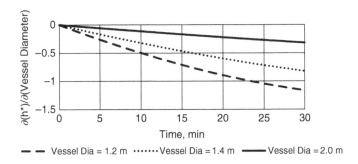

Figure 11.11   Sensitivity of water level to cylindrical vessel's diameter

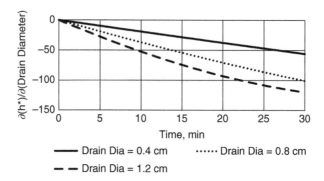

Figure 11.12   Sensitivity of water level to different drain hole diameters

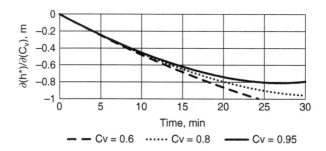

Figure 11.13   Sensitivity of water level to discharge velocity coefficient for the drain hole

The sensitivities of water levels with respect to cylindrical drain hole diameters, $\partial(h^*)/\partial(Drain\ Diameter)$, are shown as a function of time for three different drain hole diameters in Figure 11.12. The water level sensitivity to drain hole diameter increases slowly with time for small drain hole diameters.

The sensitivities of water levels with respect to discharge velocity coefficient $\partial(h^*)/\partial(C_V)$ are shown as a function of time for three different discharge velocity coefficients in Figure 11.13. For high water levels (i.e. for initial discharge times), the water level sensitivity to discharge velocity coefficient is independent of the discharge velocity coefficient value. For low water levels in the vessel, higher discharge velocity coefficients, namely less energy losses at the drain hole, decrease the water level sensitivity to discharge velocity coefficient.

12

Water Turbine Under a Dam

Water's potential energy stored behind a dam in a reservoir has been used very effectively for many decades for a spin water turbine that activates a generator to produce electricity. In this chapter we will apply the first law of thermodynamics for open systems, mostly identified as the modified Bernoulli equation, to sensitivities of design parameters for a water turbine system shown in Figure 12.1.

The Bernoulli equation for conservation of mechanical energy between location 1, namely the top level of the reservoir, and location 2, namely the tunnel entrance to the water turbine, along a streamline can be written as:

$$\frac{P_{atm}}{\rho_{water}} + g \times h = \frac{V_2^2}{2} + \frac{P_2}{\rho_{water}} \tag{12.1}$$

where the atmospheric pressure P_{atm} is assumed to be constant at the sea-level value of $1.01 \times 10^5 \, \mathrm{N\,m^{-2}}$ (or Pa), the water density is also assumed to be constant at $1000 \, \mathrm{kg\,m^{-3}}$, g is the gravitational constant with a value of $9.81 \, \mathrm{m\,s^{-2}}$, h (m) is the height from the water entrance to the turbine tunnel (i.e. point 2) to the surface of the reservoir (i.e. location 1), V_2 ($\mathrm{m\,s^{-1}}$) is the average water velocity at the tunnel entrance at location 2, and P_2 ($\mathrm{N\,m^{-2}}$) is the average pressure at the tunnel entrance at location 2 (see Figure 12.1).

Next let us analyze the energy of the water mass between the beginning of the inlet tunnel to the water turbine, location 2, and the end of the outlet tunnel from the water turbine, position 3, as shown in Figure 12.1. We can state the first law of

Case Studies in Fluid Mechanics with Sensitivities to Governing Variables, First Edition. M. Kemal Atesmen.
© 2019 John Wiley & Sons Ltd. This Work is a co-publication between John Wiley & Sons Ltd and ASME Press.

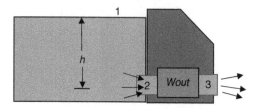

Figure 12.1 A typical water turbine system that operates on water's potential energy $\rho_{water} \times g \times h$, from a reservoir behind a dam

thermodynamics in words between locations 2 and 3 in a steady, time-averaged, and fully developed flow as follows:

> The energy of the water mass entering at location 2, namely the sum of the pressure, potential and kinetic energies, plus the net amount of heat and work added to the water mass between locations 2 and 3, minus the energy of the water mass leaving at location 3 should equal the net increase of energy in the water mass during the time interval that the water mass is between locations 2 and 3.

If we assume no net increase of energy in the water mass during its flow from location 2 to location 3, and there is no heat transfer in or out of the open system from location 2 to location 3, we can restate the first law of thermodynamics in words between locations 2 and 3 as follows:

> The energy of the water mass entering at location 2, namely the pressure, potential and kinetic energies, plus the net amount of work added to the water mass between locations 2 and 3, minus the energy of the water mass leaving at location 3 should equal zero.

We can write this first law of thermodynamics statement in equation form for a steady, time-averaged, and fully developed flow, as shown in Eq. (12.2):

$$\frac{V_2^2}{2} + \frac{P_2}{\rho_{water}} = \frac{V_3^2}{2} + \frac{P_{atm}}{\rho_{water}} + \frac{W_{out}}{\dot{m}} + H_L \qquad (12.2)$$

where W_{out} (W or N m s^{-1}) is the work done by flowing water on the turbine blades, $\dot{m} = \rho_{water} \times V_3 \times A_3$ (kg s^{-1}) is the water mass flow rate through the open system, and H_L represents head losses due to friction, elbows, minor hydrostatic variation, and so on between locations 2 and 3. Head losses are determined mainly by experiments for different surface roughness and for different ratios of inertia forces to friction forces, namely the Reynolds number, which is defined as $Re_D = \rho_{water} \times V_3 \times D/\mu$. D is the average hydraulic diameter between locations

2 and 3, defined as $D = \frac{4 \times A_{average}}{P_{average}}$. $A_{average}$ is the average area and $P_{average}$ is the average perimeter of the water tunnels between locations 2 and 3. μ is the viscosity of water for standard atmospheric conditions, namely $0.001 \, \text{N s m}^{-2}$. For the present case in hand, head losses are defined for dimensional purposes as shown in Eq. (12.3):

$$H_L = \frac{V_3^2}{2} \times \frac{L}{D} \times f \qquad (12.3)$$

where L is the distance and f the averaged friction factor between locations 2 and 3. Combining Eqs. (12.1)–(12.3) gives us the cubic equation for the unknown V_3:

$$F(V_3) = 0.5 \times \left(\frac{L}{D} \times f + 1 \right) \times V_3^3 - g \times h \times V_3 + \frac{W_{out}}{\rho_{water} \times A_3} = 0 \qquad (12.4)$$

Equation (12.4) can be solved graphically for V_3 for different potential energy heights or water heights h in the reservoir, as shown in Figure 12.2.

We can only use the higher root values of $F(V_3) = 0$, since V_3 has to increase with increasing reservoir water heights. The following parameter values are assumed for solutions in Figure 12.2:

$$\frac{L}{D} = 25, \quad f = 0.02, \quad W_{out} = 10 \, \text{MW}, \quad A_3 = 2 \, \text{m}^2$$

The average friction factor f was obtained from the experimental Moody diagram (see Ref. [9], chapter 8) for concrete tunnel walls with a relative roughness of $\frac{e}{D} = 0.001$ at high Reynolds numbers (i.e. $Re_D > 5 \times 10^7$). 10 MW of work done by flowing water on the turbine blades needs at least a reservoir height of 51.2 m in order to be a feasible design under the prescribed parameters. $V_3 = 15 \, \text{m s}^{-1}$ for this minimum potential energy condition. The exit velocity V_3 is plotted against the reservoir's potential energy height in Figure 12.3.

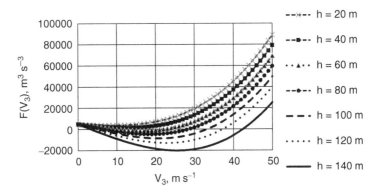

Figure 12.2 Positive roots of the function $F(V_3)$ in Eq. (12.4) for different reservoir water heights between locations 1 and 2

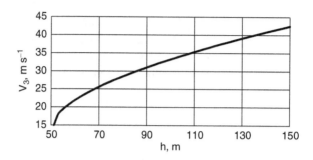

Figure 12.3 Exit velocity V_3 downstream of the turbine versus reservoir's potential energy height at the inlet to the turbine tunnel

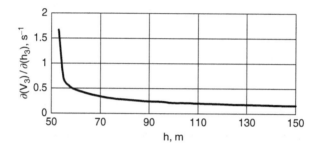

Figure 12.4 Sensitivity of exit velocity downstream of the turbine to potential energy height at the inlet tunnel to the turbine

Next let us investigate the sensitivity of V_3 to potential energy height h, as shown in Figure 12.4. Initially, as the necessary water height increases above 51.2 m between positions 2 and 1, in order to operate the turbine with the present design parameters, the exit velocity also increases fast. However, the sensitivity of the exit velocity increases to water height increases slow down at high potential energy heights. For example, for potential energy heights above 115 m, the change in exit velocity increases less than $0.2 \, \mathrm{m \, s^{-1}}$ per meter of increase in h.

Next let us analyze the product of the water tunnel geometry and the friction factor on the exit velocity, namely $(L \times f/D)$. As $(L \times f/D)$ increases, the exit velocity decreases as expected, as shown in Figure 12.5. The following parameter values are assumed for the solutions presented in Figure 12.5:

$$h = 100 \, \mathrm{m}, \quad W_{out} = 10 \, \mathrm{MW}, \quad A_3 = 2 \, \mathrm{m}^2$$

The sensitivity of the exit velocity to $(L \times f)/D$ is presented in Figure 12.6. The exit velocity V_3 decreases fast at low $(L \times f)/D$ values [i.e. $(L \times f)/D < 0.6$]. The decrease in exit velocity slows down at high $(L \times f)/D$ values [i.e. $(L \times f)/D > 0.6$].

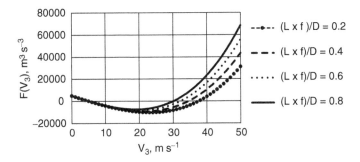

Figure 12.5 $F(V_3)$ versus V_3 for different $(L \times f)/D$ values at a reservoir water height of 100 m above position 2

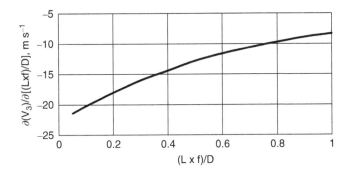

Figure 12.6 Sensitivity of exit velocity V_3 to $(L \times f)/D$

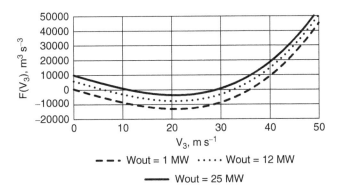

Figure 12.7 $F(V_3)$ versus V_3 for different values of work taken out of the open flow system by the turbine at a reservoir water height of 100 m between positions 1 and 2

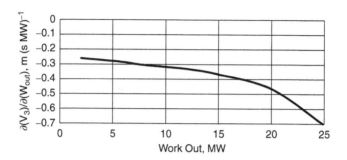

Figure 12.8 Sensitivity of exit velocity V_3 to work done on the turbine by the open flow system

Next let us analyze the behavior of the exit velocity V_3 for different values of work taken out of the open flow system by the turbine at a reservoir water height of 100 m between positions 1 and 2. The $F(V_3)$ function for several different W_{out} values are presented in Figure 12.7. As the work taken out of the open flow system increases, V_3 decreases for a fixed potential energy height. The following parameter values are assumed for the solutions presented in Figure 12.7:

$$h = 100\,\text{m}, \quad \frac{(L \times f)}{D} = 0.5, \quad A_3 = 2\,\text{m}^2$$

As the work done on the turbine by the open flow system increases, the change in the exit velocity also increases. The sensitivity of the exit velocity V_3 to work done on the turbine by the open flow system is shown in Figure 12.8.

13

Centrifugal Separation of Particles

Centrifugal acceleration forces have been used very effectively to separate solid particles from fluids or to separate different density fluids. In the medical industry, urine and blood components can be separated using centrifugal acceleration forces. For example, a normal blood solution, which has a density of approximately 1060 kg m^{-3}, can be separated into its components, because of their density gradients, by accelerating it centrifugally in a test tube. The red blood cells, which have a density of approximately 1100 kg m^{-3}, end up at the bottom of the test tube, the white blood cells, which have a density of approximately 1050 kg m^{-3}, end up on top of the red blood cells in the test tube, and the blood plasma, which has a density of approximately 1025 kg m^{-3}, ends up on top of the white blood cells in the test tube. Centrifugal acceleration forces have been used for centuries to separate milk into cream and skimmed milk, to separate yeast cells from beer, to separate solids and water in sewage sludge, to remove dust particles from the air, and so on, in different types of centrifuges.

In this chapter we will investigate the separation of particles in a fluid flow by centrifugal forces in a centrifuge shaped like a concentric cylinder. The fluid, along with spherical particles of different diameters, enters the centrifuge at the bottom and high centrifugal forces due to the rotating inner cylinder separate out the particles from the fluid. All the particles which have a critical diameter greater than D_{cr} will adhere to the walls of the stationary outer cylinder, while the fluid exits the centrifuge at the top. The exiting fluid will contain only particles with diameter smaller than D_{cr}.

Case Studies in Fluid Mechanics with Sensitivities to Governing Variables, First Edition. M. Kemal Atesmen.
© 2019 John Wiley & Sons Ltd. This Work is a co-publication between John Wiley & Sons Ltd and ASME Press.

We are going to use the drag force for a creeping flow around a spherical particle (Stokes flow), similar to that used in Chapter 5, to counter the centrifugal forces, namely a drag coefficient of $C_d = \frac{24}{Re_D}$. In our applications, in order to meet the creeping flow criterion, we have to make sure that the inertial forces are small compared to the viscous forces $\left[\text{i.e. Reynold number } Re_D = \frac{V \times D}{\left(\mu/\rho_f\right)} \ll 1\right]$. For field forces acting on a particle, we are going to use the centrifugal acceleration force instead of the gravitational force used in Chapter 5. The force balance on a particle can be written as follows:

$$(\rho_p - \rho_f) \left[\frac{4}{3}\pi(0.5D)^3\right](r\omega^2) = C_d \pi (0.5D)^2 (0.5\rho_f V^2) \tag{13.1}$$

where ρ_p (kg m^{-3}) is the particle density in the fluid, ρ_f (kg m^{-3}) is the fluid density, D (m) is the particle diameter, $r\omega^2$ (m-rad s^{-2}) is the centrifugal acceleration, C_d is the drag coefficient for creeping flow around a sphere, and V (m s^{-1}) is the radial velocity of the particle. Equation (13.1) can be rearranged to obtain the particle's radial velocity in a centrifugal force field:

$$V = \frac{dr}{dt} = \left(\frac{r\omega^2 D^2}{18\mu}\right)(\rho_p - \rho_f) \tag{13.2}$$

where r (m) is the radial distance between two concentric cylinders, the inner rotating cylinder has an outer radius of r_i (m), the outer stationary cylinder has an inner radius of r_o (m), t (s) is the time it takes for a particle to travel in the radial direction, and μ (N m s^{-2}) is the fluid's viscosity. The angular velocity ω (rad s^{-1}) can be expressed in revolutions per minute (rpm) of the centrifuge, as $\omega = \left(\frac{2\pi}{60}\right) \times N$. Then, the centrifugal acceleration can be defined in terms the Earth's gravitational acceleration (i.e. g-force):

$$r\omega^2 = 0.00111826 \times r \times N^2 \tag{13.3}$$

with r (m) and N (rpm). We can determine the radial position of a particle versus time in the centrifuge by integrating Eq. (13.2) from $r = r_i$ at $t = 0$ to $r = r$ at $t = t$, as shown:

$$r = r_i \times \exp\left[\frac{(2\pi)^2 N^2 (\rho_p - \rho_f) D^2}{60^2 \times 18 \times \mu} \times t\right] \tag{13.4}$$

If the volume flow rate through the centrifuge is given by Q (m^3 s^{-1}), then the volume of the centrifuge is $VOL = \pi \times (r_o^2 - r_i^2) \times H$ (m^3), where H (m) is the height of the centrifuge. A particle in the fluid has to reach the centrifuge's outer radius r_o at a residence time of $t_{res} = VOL/Q$. Otherwise, the particle will go out of the centrifuge with the exiting fluid. All particles larger than a critical diameter D_{cr}

will reach the outer wall and separate from the fluid. We can obtain the critical particle diameter D_{cr} from Eq. (13.4) by setting $r = r_o$ at $t = t_{res}$ as follows:

$$D_{cr} = \left[\left(\frac{1}{t_{res}} \right) \times \left(\frac{60^2}{4\pi^2 \times N^2} \right) \times \frac{(18 v_f) \times LN \left(\frac{r_o}{r_i} \right)}{\left(\frac{SPG_p}{SPG_f} - 1 \right)} \right]^{1/2} \tag{13.5}$$

where v_f (m^2 s^{-1}) is the kinematic viscosity of the fluid, SPG_p is the specific gravity of particles, and SPG_f is the specific gravity of the fluid. We can now calculate the maximum diameter of the particles that will be separated from a fluid under given centrifugal acceleration forces for a specific centrifuge shaped like a concentric cylinder. Let us assume the following parameters for our calculations:

$$r_i = 0.02 \text{ m}, r_o = 0.10 \text{ m}, H = 0.25 \text{ m}$$

The above centrifuge dimensions give a centrifuge volume of $VOL = 0.007539$ m^3. If we assume a fluid flow rate of $Q = 0.02$ m^3 s^{-1} through the centrifuge, then the residence time in the centrifuge becomes $t_{res} = 0.754$ s. Let us consider two different fluids (i.e. water and motor oil) with the following properties for our calculations:

$$SPG_{H_2O} = 1.0 \text{ and } v_{H_2O} = 1 \times 10^{-6} \text{ m}^2 \text{ s}^{-1} \text{ at } 20°C \text{ temperature}$$

$$SPG_{Motor\,Oil} = 0.9 \text{ and } v_{Motor\,Oil} = 1.22 \times 10^{-4} \text{ m}^2 \text{ s}^{-1} \text{ at } 20°C \text{ temperature}$$

The final parameter that we have to specify for our calculations is the specific gravity of the particles in the fluid, which is $SPG_p = 2.8$. The g-forces on the particles at the outer diameter of the centrifuge and the critical particle diameters (µm) in water and oil flows through this centrifuge are presented as a function of the centrifuge's rpm in Figure 13.1. As the centrifugal acceleration forces increase, the critical particle diameter decreases, namely smaller and smaller particles get stuck to the outer wall of the centrifuge. The g-forces on the particles increase as N^2, namely 11 000 g at 10 000 rpm and 280 000 g at 50 000 rpm. The critical diameters of the particles in motor oil flow are an order of magnitude larger than those in water flow due to the two orders of magnitude difference in kinematic viscosity between the two fluids, namely the motor oil flow has a two order of magnitude higher drag force than the water flow on a particle.

The smallest particle that will stay in the centrifuge and not go out of the centrifuge with the flow has path as shown in Figure 13.2. This critical diameter particle will travel from r_i to r_o in a residence time of 0.754 s.

Next let us investigate the sensitivities of the critical particle diameter to the different parameters that are used in Eq. (13.5). The sensitivity of the critical particle

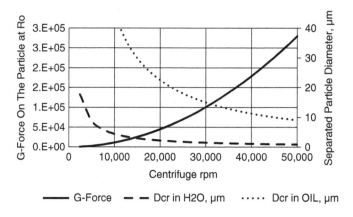

Figure 13.1   Separated critical particle diameter decreases as centrifuge rpm increases

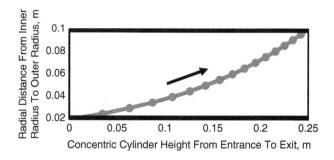

Figure 13.2   Critical diameter particle path from centrifuge entrance to exit starting at the inner radius and ending at the outer radius

diameter to changes in centrifuge rpm is given in Figure 13.3 for water and motor oil at 20°C. As the centrifuge rpm increases, the critical particle diameter decreases as N^{-1}. However, the absolute value of a change in critical particle diameter with respect to a change in centrifuge rpm, namely $\frac{\partial (D_{cr})}{\partial (N)}$, decreases as N^{-2} with increasing centrifuge rpm. The sensitivity of the critical particle diameter in motor oil is an order of magnitude higher than that in water, due to its higher kinematic viscosity.

The sensitivity of the critical particle diameter to changes in volume flow rate through the centrifuge is given in Figure 13.4 for water and motor oil at 20°C. As the volume flow rate increases, the critical particle diameter also increases as $Q^{0.5}$ because the residence time for the particles in the centrifuge decreases. However, a change in critical particle diameter with respect to a change in volume flow rate, namely $\frac{\partial (D_{cr})}{\partial (Q)}$, decreases as $Q^{-0.5}$ with increasing volume flow rate. The sensitivity

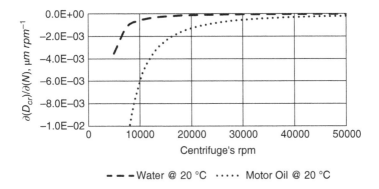

Figure 13.3   Sensitivity of critical particle diameter (μm) to change in centrifuge rpm

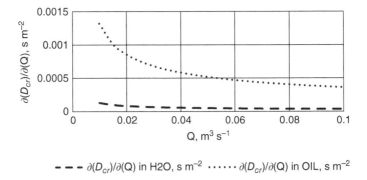

Figure 13.4   Sensitivity of critical particle diameter to change in volume flow rate through the centrifuge

of the critical particle diameter in motor oil is an order of magnitude higher than that in water, due to motor oil's higher kinematic viscosity.

The sensitivity of the critical particle diameter to changes in centrifuge volume by height H only is given in Figure 13.5 for water and motor oil at 20°C. As the volume of the centrifuge increases by height, the critical particle diameter decreases as $VOL^{-0.5}$, because the residence time of the particles in the centrifuge also increases. However, the absolute value of a change in critical particle diameter with respect to a change in volume, namely $\frac{\partial(D_{cr})}{\partial(VOL)}$, decreases as $VOL^{-1.5}$ with increasing volume. The absolute value of the sensitivity of the critical particle diameter in motor oil to changes in centrifuge volume (i.e. only by height) is an order of magnitude higher than that in water, due to motor oil's higher kinematic viscosity.

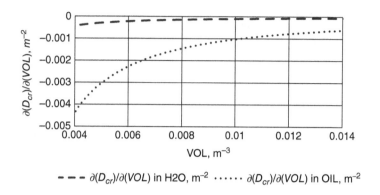

Figure 13.5 Sensitivity of critical particle diameter to change in centrifuge volume (only height changes)

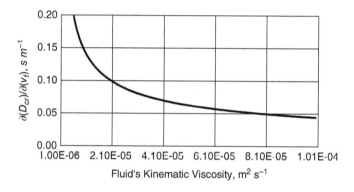

Figure 13.6 Sensitivity of critical particle diameter to change in fluid kinematic viscosity

The sensitivity of the critical particle diameter to changes in the fluid's kinematic viscosity v_f is given in Figure 13.6. As the fluid's kinematic viscosity increases, the critical particle diameter also increases as $v_f^{0.5}$. However, a change in critical particle diameter with respect to a change in the fluid's kinematic viscosity, namely $\frac{\partial(D_{cr})}{\partial(v_f)}$, decreases as $v_f^{-0.5}$ with increasing kinematic viscosity. The sensitivity of the critical particle diameter to changes in the fluid's kinematic viscosity is more prominent in low kinematic viscosities.

The sensitivity of the critical particle diameter to changes in the centrifuge's radii ratio r_o/r_i is given in Figure 13.7 for water flow. In this analysis, r_i is held constant and r_o is varied. As r_o increases, the particle's residence time increases and therefore the critical particle diameter decreases. However, the absolute value

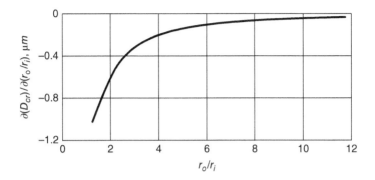

Figure 13.7   Sensitivity of critical particle diameter to change in centrifuge radii ratio r_o/r_i

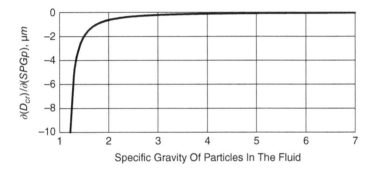

Figure 13.8   Sensitivity of critical particle diameter to particle's specific gravity

of the change in critical particle diameter with respect to a change in r_o/r_i, namely $\frac{\partial(D_{cr})}{\partial(r_o/r_i)}$, decreases with increasing r_o/r_i. The sensitivity of the critical particle diameter to changes in r_o/r_i is more prominent in low r_o/r_i values.

The sensitivity of the critical particle diameter to changes in the particle's specific gravity SPG_p is given in Figure 13.8 for water flow. As SPG_p increases, the critical particle diameter decreases. However, the absolute value of the change in critical particle diameter with respect to a change in SPG_p, namely $\frac{\partial(D_{cr})}{\partial(SPG_p)}$, decreases with increasing SPG_p. The sensitivity of the critical particle diameter to changes in SPG_p is more prominent in particle specific gravity values close to that of water, namely 1.

The sensitivity of the critical particle diameter to changes in the fluid's specific gravity SPG_f is given in Figure 13.9 for water flow at temperatures between 0 and 80°C. As the temperature of water decreases, its density and viscosity increase.

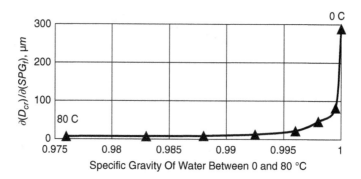

Figure 13.9   Sensitivity of critical particle diameter to water's specific gravity between 0 and 80°C

However, since water's viscosity increases faster than its density, water's specific gravity SPG_f also increases. As SPG_f increases, the critical particle diameter also increases. The change in the critical particle diameter with respect to a change in SPG_f, namely $\frac{\partial(D_{cr})}{\partial(SPG_f)}$, increases with increasing SPG_f. The sensitivity of the critical particle diameter to changes in SPG_f is more prominent at low temperatures, namely below 30°C.

14

A Simple Carburetor

The Venturi effect is the reduction in air pressure and therefore the increase in air velocity when filtered air flows through the throat section of a simple carburetor. The reduction in air pressure and therefore the increase in air velocity in the throat section of a carburetor is limited by the choked flow condition at Mach number unity. A fuel nozzle is placed in an appropriate location in the throat section to atomize the incoming fuel and mix it with air. A float in a stored fuel chamber adjusts the fuel level with respect to the throat location. Upstream of the throat area, a choke valve controls the incoming air mass flow rate to regulate a lean or rich air/fuel mixture, $A/F = \dot{m}_{air}/\dot{m}_{fuel}$, for load requirements and environmental conditions. Downstream of the throat area, a throttle valve opens and closes as actuated by the accelerator pedal, to supply the required A/F mixture to the engine inlet manifold. The ideal A/F mixture for a simple carburetor is 14.7, namely when the air flow rate-to-fuel flow rate ratio is 14.7 and the incoming fuel and oxygen will be consumed completely during the combustion process. When there is excess air flow through the simple carburetor (i.e. A/F > 18), the mixture is too lean for combustion. When there is excess fuel flow through the simple carburetor (i.e. A/F < 10), the mixture is too rich for combustion.

In the present analysis, the air mass flow rate $\dot{m}_{air}$ through a simple carburetor is assumed to have a frictionless and adiabatic flow of air in the absence of work, no change in potential energy and constant specific heat. See Chapter 15 for a detailed treatment of the first law of thermodynamics for a perfect gas in a steady and one-dimensional Venturi flow. For a perfect gas with constant specific heat

Case Studies in Fluid Mechanics with Sensitivities to Governing Variables, First Edition. M. Kemal Atesmen.
© 2019 John Wiley & Sons Ltd. This Work is a co-publication between John Wiley & Sons Ltd and ASME Press.

and temperature, the pressure of air going through the throat of the carburetor and the incoming air are related as:

$$\frac{T_{th}}{T_I} = \left(\frac{P_{th}}{P_I}\right)^{(k-1)/k} \tag{14.1}$$

where k is the ratio of the specific heat at constant pressure, $c_P = \frac{dh}{dT}$, to the specific heat for constant volume, $c_v = \frac{du}{dT}$, h is enthalpy, and u [J (kg K)$^{-1}$] is the internal energy of the flowing air. In this case, for air, k is a constant and has a value of 1.4 at 20°C. The subscript "th" in the equations stands for the carburetor throat and the subscript "I" stands for the inlet to the carburetor, where the air velocity is assumed to be negligible. With all the above assumptions, the first law of thermodynamics reduces (for details see Chapter 15) to the following:

$$h_I = h_{th} + \frac{V_{th}^2}{2} \tag{14.2}$$

where $h = u + \frac{P}{\rho} = u + R \times T$ is defined as the enthalpy of air, which is the sum of its internal energy and the specific gas constant R for air times its temperature. For a perfect gas, $R = c_P - c_v$ and for air, $R = 287.01$ m^2 (s^2 K)$^{-1}$. Using Eq. (14.2) and the definition of specific heat at constant pressure, the velocity V_{th} (m s^{-1}) at the carburetor throat can be formulated as follows:

$$V_{th} = \sqrt{2 \times c_P \times (T_I - T_{th})} \tag{14.3}$$

The air mass flow rate $\dot{m}_{air}$ (kg s^{-1}) is defined as:

$$\dot{m}_{air} = C_{th} \times \rho_{th} \times A_{th} \times V_{th} \tag{14.4}$$

where C_{th} is the coefficient of discharge for the carburetor's throat area, namely a correction factor between the theory and the real air mass flow rate through the throat. The coefficient of discharge depends on the size, shape, and friction encountered in the throat, and is assumed to have a value of 0.91 in the present analysis. ρ_{Th} (kg m^{-3}) is the density of air in the throat and A_T (m^2) is the area of the throat. The equation of state for an ideal air flow can be written, between the entry to the carburetor and its throat, as:

$$\frac{P_I}{\rho_I \times T_I} = \frac{P_{th}}{\rho_{th} \times T_{th}} \tag{14.5}$$

Combining Eqs. (14.1)–(14.5) provides the air mass flow rate through a simple carburetor:

$$\dot{m}_{air} = C_{th} \times A_{th} \times \frac{P_I}{\sqrt{R \times T_I}} \times \left(\frac{P_{th}}{P_I}\right)^{1/k}$$

$$\times \sqrt{2 \times \left(\frac{k}{k-1}\right) \times \left[1 - \left(\frac{P_{th}}{P_I}\right)^{(k-1)/k}\right]} \tag{14.6}$$

For compressible flows, the Mach number is defined as the ratio of the velocity to the local speed of sound. For the velocity of air at the carburetor's throat, the Mach number is $M = V_{th}/\sqrt{k \times R \times T_{th}}$. Now, the air mass flow rate in Eq. (14.6) can be written as a function of the throat Mach number:

$$\dot{m}_{air} = C_{th} \times A_{th} \times P_I \times \sqrt{\frac{k}{R \times T_I}} \times M$$

$$\times \left(\frac{1}{1 + 0.5 \times (k-1) \times M^2} \right)^{(k+1)/[2 \times (k-1)]} \tag{14.7}$$

The fuel mass flow rate $\dot{m}_{fuel}$ entering the throat area can be treated as an incompressible fluid using Bernoulli's equation. Assuming that the fuel level in the float chamber always stays the same as at the narrowest throat location, and the fuel exits into the carburetor from the fuel nozzle end that is z cm above the narrowest throat location, Eq. (14.8) shows the applicable mechanical energy relationship:

$$\left(\rho_{fuel} \times \frac{V_{fuel}^2}{2} \right) + P_{th} + \rho_{fuel} \times g \times z = P_I \tag{14.8}$$

The velocity of the fuel entering from the throat area into the fuel nozzle can be obtained as:

$$V_{fuel} = \sqrt{2 \times \left(\frac{P_I - P_{th}}{\rho_{fuel}} - g \times z \right)} \tag{14.9}$$

Using Eq. (14.9), the fuel mass flow rate is defined as:

$$\dot{m}_{fuel} = C_{nozzle} \times A_{nozzle} \times \sqrt{2 \times \rho_{fuel}(P_I - P_{th} - \rho_{fuel} \times g \times z)} \tag{14.10}$$

Equation (14.10) can be rewritten using the Mach number definition of $\frac{P_{th}}{P_I}$ (see Chapter 15), and Eq. (14.10) takes the following form:

$$\dot{m}_{fuel} = C_{nozzle} \times A_{nozzle} \times \sqrt{2 \times \rho_{fuel} \times P_I \times \left(1 - \frac{P_{th}}{P_I} - \frac{\rho_{fuel} \times g \times z}{P_I} \right)} \tag{14.11}$$

where C_{nozzle} is the coefficient of discharge at the fuel nozzle, namely a correction factor between the theory and the real fuel mass flow rate through the nozzle from the throat, $\frac{P_{th}}{P_I} = [1 + 0.5 \times (k-1) \times M^2]^{k/(1-k)}$. The coefficient of discharge for the fuel nozzle depends on the size, shape, and friction encountered in the nozzle, and is assumed to have a value of 0.74 in the present analysis.

Now we can analyze the air/fuel ratio (A/F) behavior in a simple carburetor by using Eqs. (14.7) and (14.11) with the following input parameters:

$$T_I = 293 \text{ K}, P_I = 101\,325 \text{ N m}^2, \frac{A_{th}}{A_{nozzle}} = 100, 300, 500$$

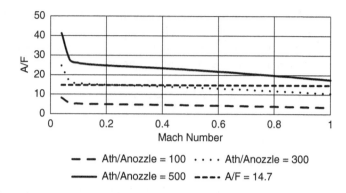

Figure 14.1   A/F versus Mach number

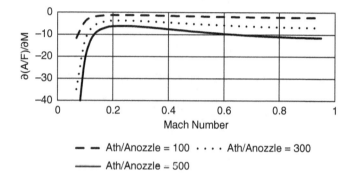

Figure 14.2   Sensitivity of A/F to Mach number

Figure 14.1 shows the air/fuel ratio as a function of Mach number for three different throat area-to-fuel nozzle area ratios. For this simple carburetor, $A_{th}/A_{nozzle} = 300$ is very close to the ideal air fuel ratio of 14.7. As A_{th}/A_{nozzle} increases, the air mass flow rate becomes excessive and the air/fuel mixture becomes too lean to burn. As A_{th}/A_{nozzle} decreases, the air mass flow rate becomes depleted and the air/fuel mixture becomes too rich to burn. The A/F ratio is very sensitive to Mach number variations at small Mach numbers (i.e. $M < 0.1$). The A/F mixture becomes leaner as the Mach number increases towards the choking value of unity.

Figure 14.2 presents the sensitivity of the air/fuel ratio to Mach number. The absolute value of this sensitivity decreases at small Mach numbers, goes through a minimum around $M = 0.2$, and starts to increase gradually until its choke value.

Figure 14.3 shows the sensitivity of P_{th}/P_I as a function of the Mach number. As the Mach number increases, the absolute value of the sensitivity also increases.

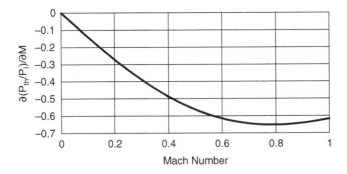

Figure 14.3   Sensitivity of P_{th}/P_I to Mach number

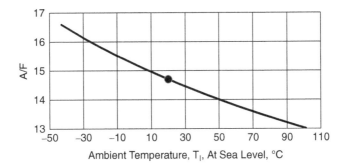

Figure 14.4   A/F versus ambient temperature T_I at sea level

The absolute value of the sensitivity goes through a maximum at $M = 0.79$ and starts to decrease gradually until P_{th}/P_I reaches the value of 0.528 at $M = 1$.

Figure 14.4 shows the variation of A/F with ambient temperature. As the ambient temperature increases, the air flow into the carburetor decreases and the air fuel mixture becomes too lean to burn. The ideal air fuel ratio of 14.7 is shown as a dot in Figure 14.4 at 20°C.

Figure 14.5 shows the sensitivity of A/F to ambient temperature at choked flow. The absolute value of the sensitivity decreases as the ambient temperature increases.

Figure 14.6 presents A/F as a function of throat diameter-to-fuel nozzle diameter ratio under the choked flow condition. A/F increases as a quadratic function of d_{th}/d_{nozzle}. At small diameter ratios the air flow is diminished and results in a rich air/fuel mixture. As the diameter ratio increases the air/fuel ratio becomes lean. In this case, a throat diameter-to-fuel nozzle diameter ratio of 20 provides the ideal burn.

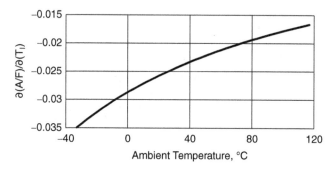

Figure 14.5   Sensitivity of A/F to ambient temperature at sea level at choked flow

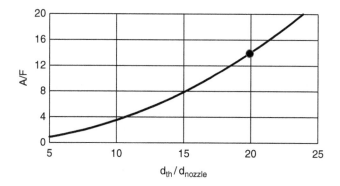

Figure 14.6   A/F versus ratio of throat diameter to fuel nozzle diameter at choked flow

Table 14.1   Effects of a $\pm 10\%$ change in nominal values of independent variables on air/fuel ratio, A/F

Parameter	Nominal value	+10% value @ $M = 0.1$	−10% value @ $M = 0.1$	+10% value @ $M = 0.5$	−10% value @ $M = 0.5$	+10% value @ $M = 1.0$	−10% value @ $M = 1.0$
d_{th}	2.00 cm	21.00	−19.00	21.00	−19.00	21.00	−19.00
d_{fuel}	1.155 mm	−17.36	23.46	−17.36	23.46	−17.36	23.46
C_{th}	0.91	10.00	−10.00	10.00	−10.00	10.00	−10.00
C_{nozzle}	0.74	−9.09	11.11	−9.09	11.11	−9.09	11.11
P_I	101 325 N m^{-2}	4.34	−4.53	4.86	−5.11	4.87	−5.12
ρ_{fuel}	737 kg m^{-3}	−4.11	4.81	−4.63	5.39	−4.65	5.40
z	1 cm	0.57	−0.56	0.02	−0.02	0.01	−0.01
k	1.4	−0.51	0.63	0.05	−0.05	0.92	−0.98
T_I	20°C	−0.34	0.34	−0.34	0.34	−0.34	0.34

When the nominal values of independent variables used in the present analysis are varied by ±10%, the resulting effects on the air/fuel ratio are as presented in Table 14.1. The resulting effects on the air/fuel ratio are shown in descending order, namely the most effective variables being the carburetor throat and fuel nozzle diameters. Next in order of importance of changing variables for A/F are the coefficients of discharge for the carburetor throat and for the fuel nozzle, whose values depend on size, shape, and friction encountered. The third tier of importance of changing variables for A/F are the atmospheric pressure and the fuel density. The least effective changing variables are the height between the narrowest throat location and the fuel exit location into the carburetor from the fuel nozzle, the ratio of the specific heats, and atmospheric temperature.

15

Ideal Gas Flow in Nozzles and Diffusers

In this chapter we will analyze properties such as temperature, pressure, and density of an ideal gas flowing through nozzles and diffusers. Analyzing the flow of compressible fluids such as air will require us to use the first and second laws of thermodynamics.

The first law of thermodynamics for a steady flow states that the net amount of energy added to a system as heat and/or work should be equal to the difference between the stored energy of the fluid mass leaving the system and the stored energy of the fluid mass entering the system. The first law can be expressed per unit mass of the fluid as, see Ref [5]:

$$q + w = (h_2 + 0.5 \times V_2^2 + g \times z_2) - (h_1 + 0.5 \times V_1^2 + g \times z_1) \tag{15.1}$$

where q is the heat added to the system, w is the work done on the system, subscript 2 represents the section where the fluid mass leaves the system, subscript 1 represents the section where the fluid mass enters the system, and h is the enthalpy of the fluid mass, defined as $h = u + \frac{P}{\rho}$ where u is the internal energy and $\frac{P}{\rho}$ is the flow work done on the system by the entering or leaving fluid mass. Next, two terms in Eq. (15.1) represent the kinetic energy and the potential energy of the fluid mass, respectively. Each term has dimension of $J\,kg^{-1}$ or equally $m^2\,s^{-2}$.

Let us now find the ideal gas properties with respect to their stagnation point properties at section 2, where the stagnation point, identified here as section s, is defined as the point at which the fluid mass is brought to rest reversibly, without

Case Studies in Fluid Mechanics with Sensitivities to Governing Variables, First Edition. M. Kemal Atesmen.
© 2019 John Wiley & Sons Ltd. This Work is a co-publication between John Wiley & Sons Ltd and ASME Press.

work, without heat transfer (i.e. adiabatically), and with negligible change in potential energy. Then, Eq. (15.1) reduces to the following:

$$0 = h_s - h_1 - 0.5 \times V_1^2 \tag{15.2}$$

For an ideal gas, stagnation enthalpy and stagnation temperature are related as:

$$h_s - h_1 = c_p \times (T_s - T_1) \tag{15.3}$$

where c_p is a thermodynamic property of a substance defined as the specific heat at constant pressure, $c_p = \left(\frac{\partial h}{\partial T}\right)_p$ [J (kg K)$^{-1}$]. Substituting Eq. (15.3) into Eq. (15.2) gives the following relationship between the ideal gas flow temperature in a nozzle or diffuser and the stagnation temperature for an ideal gas at rest:

$$\frac{T_1}{T_s} = \left(1 + \frac{0.5 \times V_1^2}{c_p \times T_1}\right)^{-1} \tag{15.4}$$

In compressible fluid mechanics, it is convenient to compare the fluid velocity to the local speed of sound in the fluid by defining the Mach number M as $M = \frac{V}{V_s}$. For ideal gases in a reversible and adiabatic flow, also called an isentropic flow, the local speed of sound is defined as $V_s = \sqrt{k \times R \times T}$ (see Ref. [6], chapter 15). k is defined as the ratio of specific heats, namely $k = \frac{c_p}{c_v}$, and the thermodynamic property of a substance defined as the specific heat at constant volume, c_v, is defined as $c_v = \left(\frac{\partial u}{\partial T}\right)_v$. Then, Eq. (15.4) can be rewritten as:

$$\frac{T_1}{T_s} = [1 + 0.5 \times (k - 1) \times M^2]^{-1} \tag{15.5}$$

For ideal gases with constant specific heats and for an isentropic process, it can be shown that $\frac{P_s}{\rho_s^k} = \frac{P_1}{\rho_1^k}$ (see Ref. [6], chapter 15). Then, the ideal gas properties with constant specific heats, for an isentropic process, can be related as:

$$\frac{T_1}{T_s} = \left(\frac{P_1}{P_s}\right)^{(k-1)/k} = \left(\frac{\rho_1}{\rho_s}\right)^{k-1} \tag{15.6}$$

Using Eqs. (15.5) and (15.6), we can obtain a relationship between the ideal gas flow pressure in a nozzle or diffuser and the stagnation pressure for an ideal gas at rest:

$$\frac{P_s}{P_1} = [1 + 0.5 \times (k - 1) \times M^2]^{k/(1-k)} \tag{15.7}$$

Using Eqs. (15.5) and (15.6), we can obtain a relationship between the ideal gas flow density in a nozzle or diffuser and the stagnation density for an ideal gas at rest:

$$\frac{\rho_1}{\rho_s} = [1 + 0.5 \times (k - 1) \times M^2]^{1/(1-k)} \tag{15.8}$$

Using Eqs. (15.4), (15.5), and (15.8), the mass flow rate per unit area, $\frac{\dot{m}_1}{A_1} = \rho_1 \times V_1$ (kg s^{-1} m^{-2}) through a nozzle or diffuser can be obtained as:

$$\frac{\dot{m}_1}{A_1} = P_s \times M \times SQRT\left(\frac{k}{R \times T_s}\right) \times [1 + 0.5 \times (k-1) \times M^2]^{(k+1)/[2\times(1-k)]}$$

(15.9)

For an ideal gas flow through a nozzle and/or diffuser, the maximum mass flow rate always occurs at $M = 1$. This maximum mass flow rate limiting condition is also called a sonic choke. For an ideal air flow with the following input parameters, the mass flow rate per unit area as a function of Mach number is shown in Figure 15.1.

Ideal air input parameters:

$$k = 1, \quad R_{air} = 286.9 \text{ m}^2(\text{s}^2 \text{ K})^{-1}$$

$$P_s = 101\,325 \text{ N m}^{-2}, \quad T_s = 293 \text{ K}$$

For the maximum mass flow rate, Eq. (15.9) reduces to the following Eq. (15.10) where the mass flow area is at a minimum at $M = 1$:

$$\left(\frac{\dot{m}_1}{A_{min}}\right)_{max} = P_s \times SQRT\left(\frac{k}{R \times T_s}\right) \times [0.5 \times (k+1)]^{(k+1)/[2\times(1-k)]}$$

(15.10)

By combining Eqs. (15.9) and (15.10), the mass flow circular area diameter ratio $\frac{d_1}{d_{min}}$ can be obtained as:

$$\frac{d_1}{d_{min}} = M^{-0.5} \times \left[\frac{1 + 0.5 \times (k-1) \times M^2}{0.5 \times (k+1)}\right]^{(k+1)/[4\times(k-1)]}$$

(15.11)

Figure 15.2 shows $\frac{d_1}{d_{min}}$ versus Mach number for an air flow. In a subsonic nozzle $M < 1$, the nozzle diameter starts to decrease to the minimum diameter as the

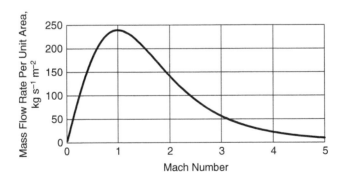

Figure 15.1   Mass flow rate per unit area versus Mach number

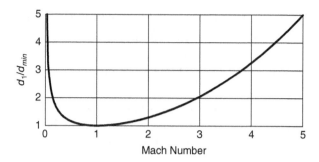

Figure 15.2   Circular area diameter versus Mach number for an isentropic flow in sonic and supersonic nozzles and diffusers

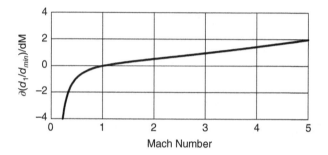

Figure 15.3   Sensitivity of d_1/d_{min} to Mach number

Mach number increases from zero to one. Above Mach number one, namely in a supersonic nozzle, the nozzle diameter increases as the Mach number increases. In a diffuser, the opposite flow phenomenon occurs with respect to the nozzle, namely in a subsonic diffuser, the diffuser diameter increases as the Mach number decreases and in a supersonic diffuser ($M > 1$), the diffuser diameter decreases as the Mach number increases.

The sensitivity of the circular area diameter to Mach number is presented in Figure 15.3. In the sonic region of a nozzle, the diameter decreases fast as the Mach number increases. The diameter decrease slows down as $M = 1$ is approached. At $M = 1$, the sensitivity goes through zero and starts to increase linearly in the supersonic region.

The ideal air properties such as temperature, pressure, and density normalized to the stagnation conditions given above are presented as isentropic flow versus Mach number in Figure 15.4. All the properties decrease as the Mach number increases, both in the sonic region and in the supersonic region for a nozzle. The decrease

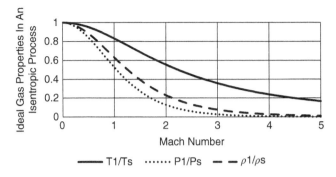

Figure 15.4   Ideal air properties normalized to given stagnation conditions shown in an isentropic flow versus Mach number

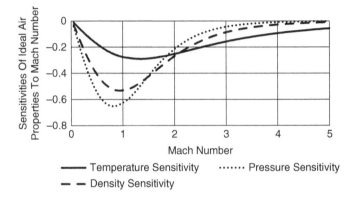

Figure 15.5   Sensitivities of ideal air properties with respect to Mach number shown in an isentropic flow versus Mach number

in pressure and density becomes negligible for Mach numbers greater than 4. At $M = 1$, $\frac{T_1}{T_s} = 0.833$, $\frac{P_1}{P_s} = 0.528$, $\frac{\rho_1}{\rho_s} = 0.634$, and $V_1 = 313.16$ m s^{-1}.

The sensitivities of the ideal air properties to Mach number are shown in Figure 15.5. The absolute values of these sensitivities increase as $M = 1$ is approached. The absolute values of these sensitivities go through a maximum and decrease to zero as the Mach number increases in the supersonic region. The pressure and density sensitivities follow each other very closely.

The ideal air velocity increases in an isentropic flow as the Mach number increases and as the fluid temperature decreases, as shown in Figure 15.6. At $M = 1$, the air velocity is 313.16 m s^{-1}. As the Mach number increases in the

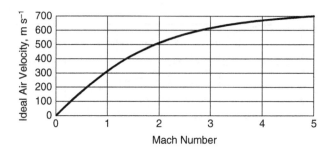

Figure 15.6   Ideal air velocity in an isentropic flow versus Mach number

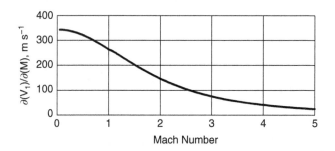

Figure 15.7   Sensitivity of ideal air velocity (m s^{-1}) to Mach number

supersonic region, the air velocity asymptotes at $767 \, \text{m s}^{-1}$. The sensitivity of the ideal air velocity to Mach number is one over the speed of sound at the local temperature. This sensitivity is presented in Figure 15.7. As the Mach number increases, the local temperature decreases and so does the sensitivity of the local velocity to the local Mach number.

16

Laminar Flow in a Pipe

In this chapter we will investigate head loss sensitivities due to friction to governing independent variables for steady, incompressible, and fully developed laminar flows through a straight pipe of constant cross-section. We will use the simplified Navier–Stokes equations in cylindrical coordinates (see Ref. [15], chapter 3) by assuming that the velocity components in the tangential and radial directions are zero. The only velocity component of the fluid flowing in the pipe, denoted by U, is in the axial direction, is constant, and only varies in the radial direction r. Also, the pressure gradient along the axial direction of the pipe, $\frac{dP}{dx}$, is a constant. With the above assumptions, the Navier–Stokes equations in cylindrical coordinates simplify to:

$$0 = -\frac{dP}{dx} + \mu \times \left(\frac{d^2U}{dr^2} + \frac{1}{r} \times \frac{dU}{dr} \right) \tag{16.1}$$

where μ ($\mathrm{N\,s\,m^{-2}}$) is the coefficient of viscosity for the fluid. Since the velocity components in the tangential and radial directions are zero, and since the velocity component in the axial direction is a constant, the equation of continuity in cylindrical coordinates is always satisfied for the present problem, namely for a steady, incompressible, and fully developed laminar flow through a straight pipe of constant cross-section. The differential equation in Eq. (16.1) can be solved by applying the boundary condition $U = 0$ at $r = R$, where R (m) is the interior radius of the pipe.

In order to be able to solve the differential equation in Eq. (16.1), let us introduce a new variable $S = \frac{dU}{dr}$. Then, Eq. (16.1) becomes a first-order differential equation as follows:

$$\frac{dS}{dr} + \frac{S}{r} = -\frac{1}{\mu} \times \frac{dP}{dx} \tag{16.2}$$

Case Studies in Fluid Mechanics with Sensitivities to Governing Variables, First Edition. M. Kemal Atesmen.
© 2019 John Wiley & Sons Ltd. This Work is a co-publication between John Wiley & Sons Ltd and ASME Press.

The complementary solution S_c to the above differential equation is:

$$S_c = \frac{C_1}{r} \tag{16.3}$$

The particular solution S_p to the differential equation in Eq. (16.2) is given as:

$$S_p = C_2 \times r \tag{16.4}$$

where C_2 can be obtained from Eq. (16.2) and is given as:

$$C_2 = -\frac{1}{2\mu} \times \frac{dP}{dx} \tag{16.5}$$

The solution to the differential equation in Eq. (16.2) is the sum of the homogeneous and particular solutions, given as:

$$S = -\frac{1}{2\mu} \times \frac{dP}{dx} \times r + \frac{C_1}{r} \tag{16.6}$$

The differential equation in Eq. (16.6) has to be integrated one more time to obtain the velocity distribution in a pipe:

$$U = -\frac{1}{4\mu} \times \frac{dP}{dx} \times r^2 + C_1 \times \ln r + C_3 \tag{16.7}$$

In Eq. (16.7), $C_1 = 0$, since the axial velocity is finite at the center of the pipe. C_3 can be obtained from the boundary condition at the pipe wall given above, namely $U = 0$ at $r = R$. C_3 is given as:

$$C_3 = \frac{1}{4\mu} \times \frac{dP}{dx} \times R^2 \tag{16.8}$$

The velocity distribution in Eq. (16.7) can be rewritten using the above constant values in final form:

$$U = \frac{1}{4\mu} \times \frac{dP}{dx} \times (R^2 - r^2) \tag{16.9}$$

We see that the fluid flowing in the pipe has a paraboloid velocity profile, namely the rotation of the parabola given by Eq. (16.9) around the centerline of the pipe. According to experiments, Eq. (16.9) is applicable for Reynolds numbers up to 2300, namely $Re_D \leq 2300$, where $Re_D = \frac{U_{ave} \times D}{v}$. U_{ave} (m s^{-1}) is the average velocity at the cross-section of the pipe, D (m) is the inner diameter of the pipe, and v (m^2 s^{-1}) is the kinematic viscosity of the fluid. For flows with Reynolds number above 2300, the flow characteristics become turbulent and the above equations no longer apply. The maximum velocity occurs at the centerline of the pipe at $r = 0$, given as:

$$U_{max} = \frac{1}{4\mu} \times \frac{dP}{dx} \times R^2 \tag{16.10}$$

For flow analysis in pipes, mostly the volume flow rate Q is used, where $Q = U_{ave} \times \pi \times R^2$ (m^3 s^{-1}). The average velocity is obtained by integrating the velocity profile in Eq. (16.9) across the radial cross-section of the pipe and then dividing the result by the cross-sectional area of the pipe, namely obtaining the first moment of the velocity distribution, as shown in Eqs. (16.11) and (16.12):

$$U_{ave} = \left(\frac{1}{\pi R^2}\right)\left(\frac{1}{4\mu} \times \frac{dP}{dx}\right) \int_0^R (R^2 - r^2) \times 2\pi r dr \qquad (16.11)$$

or

$$U_{ave} = \left(\frac{1}{8\mu} \times \frac{dP}{dx}\right) R^2 = \frac{U_{max}}{2} \qquad (16.12)$$

The pressure drop in pipe flow is attributed to the friction between the fluid and the pipe's walls, as can be seen from the wall shear stress:

$$\tau = \mu \times \left(\frac{dU}{dr}\right)_{r=R} = -\frac{R}{2} \times \frac{dP}{dx} \qquad (16.13)$$

Also, the head loss due to friction in a pipe is defined by the pressure drop in pipe flow as shown in Eq. (16.14). With the aid of Eq. (16.12), the head loss due to friction for laminar flow in a pipe can be written as:

$$h_l = \frac{\Delta P}{\rho \times g} = \frac{32 \times v \times L \times U_{ave}}{g \times D^2} \qquad (16.14)$$

where L (m) is the length of the pipe along which the pressure drop occurs. In engineering practice, head loss is also expressed in terms of a friction factor f, so that the time-averaged experimental results can be applied to Bernoulli's equation. This new definition of the friction factor provides us with a relationship between the pressure gradient in the pipe and the average flow velocity. The definition of head loss in terms of friction factor f is given by:

$$h_l = \left(\frac{U_{ave}^2}{2 \times g}\right) \times \left(\frac{L}{D}\right) \times f \qquad (16.15)$$

If we combine Eqs. (16.14) and (16.15), we obtain the following friction factor and Reynolds number relationship for laminar flow in a pipe:

$$f = \frac{64}{Re_D} \qquad (16.16)$$

In engineering practice, mostly the volume flow rate is used instead of the average velocity when calculating the head loss due to friction. Equation (16.14) can then be rewritten as a function of volume flow rate:

$$h_l = \frac{128 \times v \times L \times Q}{\pi \times g \times D^4} \qquad (16.17)$$

Now let us investigate the sensitivities of head loss due to friction to governing independent variables using Eq. (16.17). The head loss due to friction is linearly related to the fluid's kinematic viscosity, the pipe's length, and the fluid's volume flow rate. Fluids with different viscosities, densities, and kinematic viscosities are shown in Table 16.1. As the kinematic viscosity of a fluid increases or decreases with a certain percentage, so does the head loss due to friction increase or decrease with the same percentage, as long as the other independent variables stay the same. Similar arguments apply to changes in pipe length and fluid volume flow rate. The results shown in Figures 16.1–16.3 assume a pipe length of 1000 m and a fluid volume flow rate of 0.0005 m^3 s^{-1}.

The pipe's inner diameter is another story. The sensitivity of head loss due to friction varies with the pipe's inner diameter as $-D^{-5}$. This sensitivity is shown in Figure 16.1 for three different fluids. The change in head loss due to friction as the inner diameter of the pipe changes at small inner pipe diameters is substantial. This sensitivity approaches zero for low kinematic viscosity fluids such as water and gasoline at 20°C for inner pipe diameters greater than 0.2 m. Also, this sensitivity

Table 16.1 Viscosity, density, and kinematic viscosity of different fluids at different temperatures

Fluid	Viscosity $(N\,s\,m^{-2})$	Density $(kg\,m^{-3})$	Kinematic viscosity $(m^2\ s^{-1})$
Glycerine @ 20°C	1.5	1259	1.19×10^{-3}
Motor oil @ 0°C	1.1×10^{-1}	892	1.23×10^{-4}
Motor oil @ 20°C	3.0×10^{-2}	880	3.41×10^{-5}
Water @ 20°C	1.0×10^{-3}	1000	1.00×10^{-6}
Water @ 90°C	3.2×10^{-4}	965	3.32×10^{-7}
Gasoline @ 20°C	2.9×10^{-4}	737	3.93×10^{-7}
Air @ 20°C and 1 atm	1.8×10^{-5}	1.2	1.50×10^{-5}

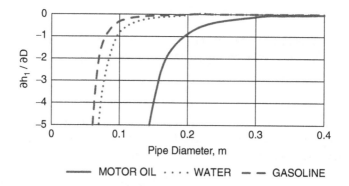

Figure 16.1 Sensitivity of head loss to pipe diameter as a function of pipe diameter for three different fluids at 20°C

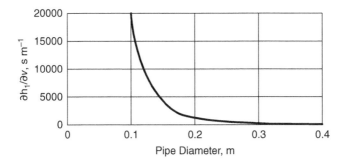

Figure 16.2 Sensitivity of head loss to kinematic viscosity as a function of pipe diameter

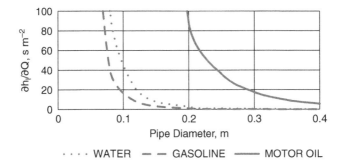

Figure 16.3 Sensitivity of head loss to volume flow rate as a function of pipe diameter for three different fluids at 20°C

approaches zero for high kinematic viscosity fluids such as motor oil at 20°C for inner pipe diameters greater than 0.3 m.

The sensitivity of head loss due to friction varies with the fluid's kinematic viscosity with respect to the pipe's inner diameter as D^{-4}. This sensitivity is shown in Figure 16.2 and applies to all fluids. The change in head loss due to friction as the fluid's kinematic viscosity changes at small inner pipe diameters is substantial. This sensitivity approaches zero for all fluids for inner pipe diameters greater than 0.3 m.

The sensitivity of head loss due to friction also varies with the fluid's volume flow rate with respect to the pipe's inner diameter as D^{-4}. This sensitivity is shown in Figure 16.3 for three fluids with different kinematic viscosities. The changes in head loss due to friction as the fluid's volume flow rate changes at small inner pipe diameters is substantial. This sensitivity approaches zero for low kinematic viscosity fluids such as water and gasoline at 20°C for inner pipe diameters greater than 0.2 m. Also, this sensitivity approaches zero for high kinematic viscosity fluids such as motor oil at 20°C for inner pipe diameters greater than 0.4 m.

17

Water Supply from a Lake to a Factory

In this chapter we will investigate the power input requirements for a water supply line from a lake to a factory for different volume flow rates and different pipe internal diameters. Water flow rates are in the turbulent zone, namely Reynolds number greater than 2300. For piping, commercial steel pipes are used with different roughness, depending on the pipe's internal diameter. Water properties in the pipe are considered as constants. The first law of thermodynamics can be integrated over the cross-section of the pipe under steady-state conditions to provide the following Eq. (17.1) between the inlet and outlet of the pipe (see Ref. [16], chapter 5):

$$\frac{U_i^2}{2 \times g} + \frac{P_i}{\rho \times g} + Z_i + \frac{W_{pump}}{\rho \times Q \times g} = \frac{U_o^2}{2 \times g} + \frac{P_o}{\rho \times g} + Z_o + h_{l-friction} + h_{l-other}$$

$$(17.1)$$

where U_i ($\mathrm{m\,s^{-1}}$) is the average water velocity entering the pipe system, g is the gravitational acceleration at sea level, namely $9.81\,\mathrm{m\,s^{-2}}$, P_1 ($\mathrm{N\,m^{-2}}$) is the water pressure entering the pipe system, ρ is the water density at 20°C, namely $1000\,\mathrm{kg\,m^{-3}}$, Z_i (m) is the potential energy elevation of the pipe system inlet, W_{pump} ($\mathrm{N\,m\,s^{-1}}$) is the pump power input required to give us the desired water volume flow rate, Q ($\mathrm{m^3\,s^{-1}}$) is the required water volume flow rate, U_o ($\mathrm{m\,s^{-1}}$) is the average water velocity exiting the pipe system, P_o ($\mathrm{N\,m^{-2}}$) is the water pressure exiting the pipe system, Z_o (m) is the potential energy elevation of the pipe system outlet, $h_{l-friction}$ (m) is the total head loss in the pipe system due to friction, namely the shear stress at the pipe's walls, and $h_{l-other}$ (m) is the total sum of head

Case Studies in Fluid Mechanics with Sensitivities to Governing Variables, First Edition. M. Kemal Atesmen.
© 2019 John Wiley & Sons Ltd. This Work is a co-publication between John Wiley & Sons Ltd and ASME Press.

losses due to bends, valves, couplings, elbows, water entrance and exit, and so on that cause a pressure drop in the pipe system. Equation (17.1) is also called the modified Bernoulli equation.

Next, the Bernoulli equation has to be written between the surface of a large lake and the inlet to the pipe system below the surface of the lake, as shown:

$$\frac{P_{atm}}{\rho \times g} + H = \frac{U_i^2}{2 \times g} + \frac{P_i}{\rho \times g} \tag{17.2}$$

where P_{atm} is the atmospheric pressure at sea level, namely $101\,325\,\mathrm{N\,m^{-2}}$, and H is the height from the inlet of the pipe system to the surface of a large lake, assumed to be $50\,\mathrm{m}$.

Using Eqs. (17.1) and (17.2), the pump power input requirements can be calculated for different water volume flow rates and different inner diameter pipes by making the following assumptions and by finding the head losses due to friction in the pipe and due to head losses in other components.

Assumptions:

$$Z_i = Z_o, P_o = 506\,625\,\mathrm{N\,m^{-2}}, \text{ total pipe length } L = 5000\,\mathrm{m}$$

water kinematic viscosity @ $20°C$: $v = 1 \times 10^{-6}\,\mathrm{m^2\,s^{-1}}$

The head losses due to friction are defined in terms of a dimensionless friction factor f:

$$h_{l-friction} = f \times \frac{L}{D} \times \frac{U_o^2}{2 \times g} = f \times \frac{8}{\pi^2} \times \frac{L}{D^5} \times \frac{Q^2}{g} \tag{17.3}$$

The friction factors are obtained experimentally for the turbulent flow regions for pipes with different internal wall roughness. These experimental results are provided in most fluid mechanics books as Moody diagrams, as a function of Reynolds number $Re_D = \frac{U_{ave} \times D}{v} = \frac{4}{\pi} \times \frac{Q}{v \times D}$ and pipe wall roughness (see Ref. [15], chapter 20). The pipe wall roughness data is obtained from the manufacturer for a certain pipe material and a certain pipe diameter.

$h_{l-other}$, the total sum of head losses due to bends, valves, couplings, elbows, water entrance and exit, can be obtained from the manufacturer for each component. These head losses are obtained experimentally, and the results depend on the Reynolds number and the component's diameter. The total sum of other head loss coefficients, K_i, is given by:

$$h_{l-other} = \left(\sum_{i=1}^{N} K_i \right) \times \frac{U_o^2}{2 \times g} = \left(\sum_{i=1}^{N} K_i \right) \times \frac{8}{\pi^2} \times \frac{1}{D^4} \times \frac{Q^2}{g} \tag{17.4}$$

The total sum of other head losses can be obtained from different manufacturers' experimental data. The total sum of other head losses for the present calculations is assumed to be 398 for a Reynolds number of $127\,000$ and a pipe internal diameter of $0.1\,\mathrm{m}$.

Using Eqs. (17.1)–(17.4) with Moody's friction factor diagrams and manufacturer's head loss coefficient data for other components, calculations were performed for the following volume flow rate and pipe internal diameter matrix:

$$Q = 0.1, 0.2, 0.3, 0.4, 0.5, \text{ and } 0.6 \text{ m}^3 \text{ s}^{-1}$$

$$D = 0.1, 0.15, 0.2, 0.25, \text{ and } 0.3 \text{ m}$$

The required pump power for the present water system versus volume flow rate for different pipe internal diameters is shown in Figure 17.1. Small-diameter pipes need much more power to balance friction losses and other losses, as expected.

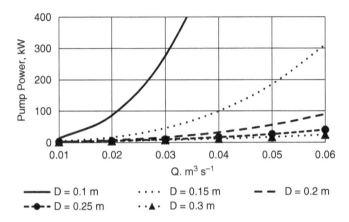

Figure 17.1 Required pump power for water flow versus volume flow rate with different pipe internal diameters

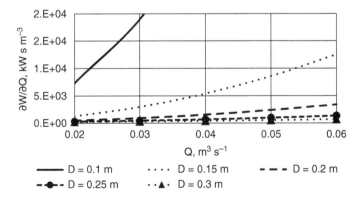

Figure 17.2 Sensitivity of required pump power to volume flow rate for different pipe internal diameters

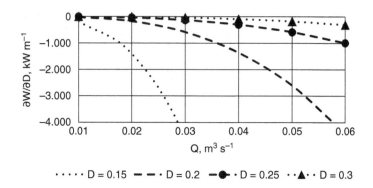

Figure 17.3 Sensitivity of required pump power to pipe internal diameter for different pipe internal diameters

As the pipe diameter increases, the pump power requirements decrease fast. However, the cost of the piping system will also increase as the pipe diameter gets larger. So, a cost feasibility analysis has to be made in order to find the pipe diameter that will provide the minimum installation and maintenance costs for the present water system.

The sensitivity of the required pump power to volume flow rate for different pipe internal diameters is shown in Figure 17.2. The change in required pump power for an increasing change in volume flow rate is always positive, as expected. This sensitivity is very prominent for pipes with small internal diameters. As the pipe internal diameter increases, this sensitivity also decreases.

The sensitivity of the required pump power to pipe internal diameter for different pipe internal diameters is shown in Figure 17.3. The change in required pump power for an increasing change pipe internal diameter is always negative, as expected. This negative sensitivity is very prominent for pipes with small internal diameter. As the pipe internal diameter increases, this negative sensitivity approaches zero.

18

Air or Water Flow Required to Cool a PC Board

In this chapter we will investigate the air and water flow required to cool a printed circuit (PC) board. The components of a PC board generate power, and these components have to be kept at a specified operating temperature. The power generation of these components depends upon the available cooling flow under the PC board. The thermal management of a PC board can be very tricky. A simple PC board is composed of three layers (i.e. a PC board attached by a thin layer of epoxy to an aluminum base plate acting as a heat sink for a uniform conduction thermal path). Let us consider a PC board whose length and width dimensions are 0.25 m by 0.25 m, namely a board area of $A = 0.0625$ m^2. The components of the PC board have to be kept at a maximum of 50°C. The cooling fluid is flowing under the aluminum base plate as a laminar flow and/or a turbulent flow over a flat surface. In most PC board designs, about 90% of the heat generated is extracted by the flowing fluid through convective heat transfer. We will consider the fluid flow to be steady, incompressible with constant properties.

Assume that the power generated by the PC board components has to equal the heat loss through a series circuit heat transfer resistances between PC components and the flowing fluid, see Ref. [1]:

$$Q = \frac{(T_{component} - T_{fluid})}{\sum_{i=1}^{4} R_i} \tag{18.1}$$

Case Studies in Fluid Mechanics with Sensitivities to Governing Variables, First Edition. M. Kemal Atesmen.
© 2019 John Wiley & Sons Ltd. This Work is a co-publication between John Wiley & Sons Ltd and ASME Press.

Here, R_1 is the conduction heat transfer resistance for the PC board:

$$R_1 = \frac{t_{PCB}}{A \times k_{PCB}} = \frac{0.005 \text{ m}}{0.0625 \text{ m}^2 \times 8.0 \text{ W (m K)}^{-1}} = 0.01 \text{ K W}^{-1} \qquad (18.2)$$

R_2 is the conduction heat transfer resistance for the epoxy layer:

$$R_2 = \frac{t_{EPOXY}}{A \times k_{EPOXY}} = \frac{0.0006 \text{ m}}{0.0625 \text{ m}^2 \times 12.0 \text{ W (m K)}^{-1}} = 0.0008 \text{ K W}^{-1} \qquad (18.3)$$

R_3 is the conduction heat transfer resistance for the aluminum heat sink back plate:

$$R_3 = \frac{t_{ALUMINUM}}{A \times k_{ALUMINUM}} = \frac{0.006 \text{ m}}{0.0625 \text{ m}^2 \times 205.0 \text{ W (m K)}^{-1}} = 0.0005 \text{ K W}^{-1} \qquad (18.4)$$

R_4 is the plate length averaged forced convection heat transfer resistance between the aluminum back plate and the cooling fluid:

$$R_4 = \frac{1}{A \times h_{ave}} \qquad (18.5)$$

where h_{ave} [W (m^2 K)$^{-1}$] is the plate length averaged forced convection heat transfer coefficient between the surface of the aluminum plate and the boundary layer of the flowing fluid. We know all the variables to solve the PC board cooling problem, except h_{ave}.

The forced convection heat transfer coefficient is determined by making an analogy between the wall shear stress and the rate of heat flow per unit area perpendicular to the flat plate surface. In the literature, this is called the Reynolds analogy (see Ref. [15], chapters 14 and 19). The Reynolds analogy has been modified with the results of experiments, and the modified version of the Reynolds analogy is given in Eq. (18.6):

$$\frac{Nu_x}{Re_x \times Pr^{1/3}} = 0.5 \times C_{fx} \qquad (18.6)$$

where x (m) is the distance from the leading edge of the flat plate, $Nu_x = \frac{(h_x) \times (x)}{k}$ is the local Nusselt number, h_x [W (m^2 K)$^{-1}$] is the local forced convection heat transfer coefficient at the surface of the flat plate, and k [W (m K)$^{-1}$] is the thermal conductivity of flowing fluid. $Re_x = \frac{(U_\infty) \times (x)}{v}$ is the local Reynolds number, U_∞ (m s^{-1}) is the fluid velocity outside the boundary layer, and v (m^2 s^{-1}) is the fluid's kinematic viscosity. $Pr = \frac{v}{\alpha}$ is the Prandtl number, defined as the ratio of momentum molecular diffusivity to thermal molecular diffusivity, where $\alpha = \frac{k}{\rho \times c_p}$, ρ (kg m^{-3}) is the density of flowing fluid, and c_p [(W s) (kg K)$^{-1}$] is the specific heat of flowing fluid at constant pressure. For air, $Pr = 0.7$ and for water, $Pr = 7.0$ at 25°C and atmospheric pressure. $C_{fx} = \frac{2 \times \tau_{0x}}{\rho \times U_\infty^2}$ is defined as the local skin friction coefficient, where τ_{0x} ($N\,m^{-2}$) is the local shear stress at the surface of the flat plate. The local skin friction coefficient at the surface of a flat plate for a laminar boundary layer has been determined exactly and is given in Eq. (18.7) (see Ref. [15], chapter 7):

$$C_{fx-laminar} = 0.664 \times Re_x^{-0.5} \qquad (18.7)$$

By combining Eqs. (18.6) and (18.7), we can obtain the local forced convection heat transfer coefficient for a laminar flow, namely $Re_x < 500\,000$, along the surface of a flat plate:

$$h_{x-laminar} = 0.332 \times Pr^{1/3} \times \left(\frac{k}{x}\right) \times Re_x^{1/2} \qquad (18.8)$$

If the flow along a flat plate is totally in the laminar boundary layer region, we can obtain a plate length averaged forced convection heat transfer coefficient as shown:

$$h_{ave-laminar} = \left(\frac{1}{L}\right) \int_0^L h_x \times dx = 0.664 \times Pr^{1/3} \times \left(\frac{k}{L}\right) \times Re_L^{1/2} \qquad (18.9)$$

The Reynolds analogy in Eq. (18.6) can be extended to the turbulent boundary layer flow along a flat plate by assuming that the molecular diffusivity and time averaged turbulent diffusivity, which is called the eddy diffusivity in the literature (see Ref. [9], chapter 6), are additive. The local skin friction coefficient at the surface of a flat plate for a turbulent boundary layer has been determined by experiments, and is given in Eq. (18.10) (see Ref. [9], chapter 6):

$$C_{fx-turbulent} = 0.0576 \times Re_x^{-0.2} \qquad (18.10)$$

Equation (18.10) is valid for a turbulent boundary layer flow along a flat plate when $1 \times 10^5 < Re_x < 1 \times 10^7$. Combining Eqs. (18.6) and (18.10) provides the local forced convection heat transfer coefficient for a turbulent flow along the surface of a flat plate:

$$h_{x-turbulent} = 0.0288 \times Pr^{1/3} \times \left(\frac{k}{x}\right) \times Re_x^{4/5} \qquad (18.11)$$

If the flow along a flat plate is totally in the turbulent boundary layer region, we can obtain a plate length averaged forced convection heat transfer coefficient as shown:

$$h_{ave-turbulent} = \left(\frac{1}{L}\right) \int_0^L h_x \times dx = 0.036 \times Pr^{1/3} \times \left(\frac{k}{L}\right) \times Re_L^{4/5} \qquad (18.12)$$

Most often, the transition from laminar boundary layer to turbulent boundary layer occurs over the flat plate. This type of transition occurs in a region, but in this analysis let us assume that the transition occurs at a point on the plate, say at length $L_{transition} < L$. In such a case, we can obtain a plate length averaged forced convection heat transfer coefficient by integrating the laminar boundary layer contribution, Eq. (18.8), from the leading edge of the plate to the transition point and adding it to the contribution from the turbulent boundary layer by integrating Eq. (18.11) from the transition point to the end of the flat plate. The plate length averaged forced convection heat transfer coefficient for this laminar plus turbulent mixed boundary layer case is shown in Eq. (18.13):

$$h_{ave-mixed\,bl} = \left(\frac{1}{L}\right)\left[\int_0^{L_{transition}} h_{x-laminar}dx + \int_{L_{transition}}^{L} h_{x-turbulent}dx\right]$$

$$h_{ave-mixed\,bl} = Pr^{1/3}\left(\frac{k}{L}\right)\left[0.664Re_{L_{transition}}^{1/2} - 0.036\left(Re_{L_{transition}}^{4/5} - Re_L^{4/5}\right)\right]$$

$$(18.13)$$

Now we can analyze the air and water flow required to cool the PC board described above by using the following parameters:

$$T_{component} = 50°C, \; T_{fluid} = 25°C, \; Re_{L_{transition}} = 500\,000$$

$$v_{air} = 1.5 \times 10^{-5}\,m^2\,s^{-1}, \; Pr_{air} = 0.7, \; k_{air} = 0.026\,W\,(m\,K)^{-1}$$

$$v_{water} = 1.0 \times 10^{-6}\,m^2\,s^{-1}, \; Pr_{water} = 7.0, \; k_{water} = 0.598\,W\,(m\,K)^{-1}$$

Let us first analyze the air flow required to cool the PC board described above using Eqs. (18.1)–(18.5), (18.9), and (18.13) with the given parameters. Figure 18.1 shows the plate length averaged forced convection heat transfer coefficient as a function of Reynolds number in the laminar boundary layer region up to $Re_L \leq 5 \times 10^5$ and in the laminar plus turbulent boundary layers region for $5 \times 10^5 < Re_L \leq 5 \times 10^6$. We should pay attention to the fact that the air flow is in the subsonic region, namely at sea level the air velocity has to be less than $340\,m\,s^{-1}$, since we have assumed the fluid flow to be incompressible with constant properties. The plate length averaged forced convection heat transfer coefficient increases much faster in the turbulent boundary layer region compared

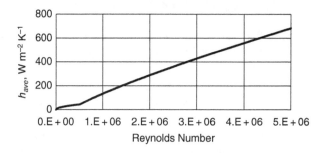

Figure 18.1 Plate length averaged h_{ave} versus Reynolds number in air flow

to the laminar boundary layer region, namely $Re_L^{0.8}$ increases in the turbulent region, see Eq. (18.12), versus $Re_L^{0.5}$ increases in the laminar region, see Eq. (18.9).

The sensitivity of the plate length averaged forced convection heat transfer coefficient to Reynolds number as a function of Reynolds number is presented in Figure 18.2. The plate length averaged forced convection heat transfer coefficient sensitivity to Reynolds number decreases much faster in the laminar boundary layer region compared to the turbulent boundary layer region, namely $Re_L^{-0.5}$ decreases in the laminar region versus $Re_L^{-0.2}$ decreases in the turbulent region. The discontinuity between the two curves occurs at $Re_L = 1 \times 10^5$.

We now have all the variables determined to solve Eq. (18.1) for possible power generation by PC components in order to keep them at a temperature of 50°C depending upon the air flow rate. The possible power generation by PC components versus Reynolds number in air flow is shown in Figure 18.3. If we want to keep in the laminar flow region throughout the whole plate, the PC components can generate a maximum of 66 W. However, when we increase the speed of air flow to $300\,\mathrm{m\,s^{-1}}$ (i.e. $Re_L = 5 \times 10^6$), the PC components can generate a power of 720 W.

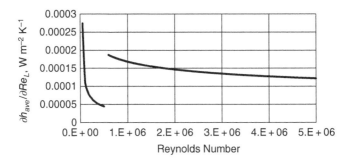

Figure 18.2 Sensitivity of plate length averaged forced convection heat transfer coefficient to Reynolds number as a function of Reynolds number in air flow

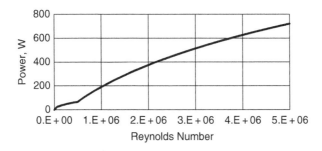

Figure 18.3 Possible power generation by PC components versus Reynolds number in air flow

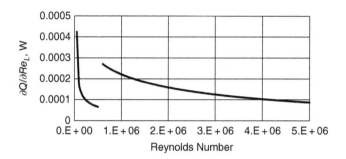

Figure 18.4 Sensitivity of power generation to Reynolds number as a function of Reynolds number in air flow

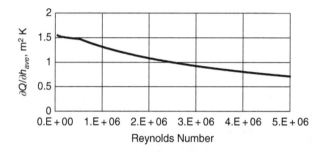

Figure 18.5 Sensitivity of power generation to plate length averaged forced convection heat transfer coefficient as a function of Reynolds number in air flow

The sensitivity of power generation by PC components to Reynolds number as a function of Reynolds number is given in Figure 18.4. This sensitivity behavior is similar to the sensitivity of the plate length averaged forced convection heat transfer coefficient to Reynolds number presented in Figure 18.2, as expected.

The sensitivity of power generation to plate length averaged forced convection heat transfer coefficient as a function of Reynolds number in air flow is shown in Figure 18.5. This sensitivity is more prominent in the laminar region than in the turbulent region. As the Reynolds number increases, this sensitivity decreases both in the laminar and in the turbulent regions.

Next, let us analyze the PC board cooling with water flowing under the aluminum plate. Figure 18.6 shows the plate length averaged forced convection heat transfer coefficient as a function of Reynolds number in the laminar boundary layer region up to $Re_L \leq 5 \times 10^5$ and in the laminar plus turbulent boundary layer region for $5 \times 10^5 < Re_L \leq 5 \times 10^6$ using Eqs. (18.9) and (18.13). The plate

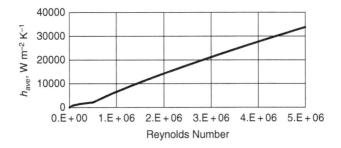

Figure 18.6   Plate length averaged h_{ave} versus Reynolds number in water flow

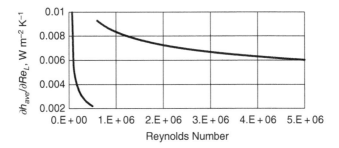

Figure 18.7   Sensitivity of plate length averaged forced convection heat transfer coefficient to Reynolds number as a function of Reynolds number in water flow

length averaged forced convection heat transfer coefficients for water flow are 50 times larger than those for air flow due to the fluid property differences: $(k * Pr^{0.333})_{water}/(k * Pr^{0.333})_{air} = 49.51$.

The sensitivity of the plate length averaged forced convection heat transfer coefficient to Reynolds number as a function of Reynolds number is presented in Figure 18.7 for water flow. This sensitivity of the plate length averaged forced convection heat transfer coefficient to Reynolds number is 50 times larger for water flow than for air flow.

The possible power generation by PC components for water flowing in the laminar region is 20-fold higher than for the air flow. As the Reynolds number increases and the laminar plus turbulent regions are encountered, the power generation asymptotes because the conduction heat transfer resistance limits are approached by the present PC board structure. The plate length averaged forced convection heat transfer resistance approaches zero. This behavior of the possible power generation in water flow is shown in Figure 18.8.

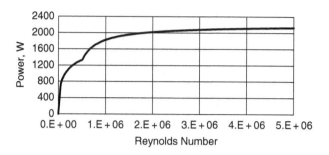

Figure 18.8 Possible power generation by PC components versus Reynolds number in water flow

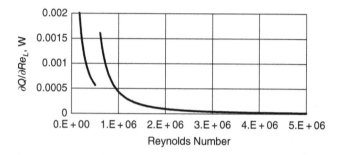

Figure 18.9 Sensitivity of power generation to Reynolds number as a function of Reynolds number in water flow

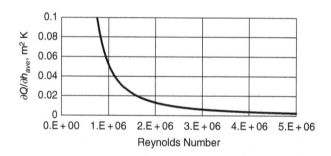

Figure 18.10 Sensitivity of power generation to average forced convection heat transfer coefficient as a function of Reynolds number in water flow

The sensitivity of power generation to plate length averaged forced convection heat transfer coefficient as a function of Reynolds number in water flow is shown in Figure 18.9. This sensitivity is more prominent in the laminar region than in the turbulent region. As the Reynolds number increases, this sensitivity decreases and asymptotes to zero as the conduction heat transfer resistance value is approached.

The sensitivity of power generation to the plate length averaged forced convection heat transfer coefficient as a function of Reynolds number in water flow is shown in Figure 18.10. This sensitivity is more prominent in the laminar region than in the turbulent region. When the Reynolds number increases, this sensitivity approaches zero as the power generation is limited by the conduction heat transfer resistance.

19

Convection Mass Transfer Through Air–Water Interface

In this chapter we will investigate the rate of mass convection (i.e. evaporation flux) from a flat water surface to air that is flowing over the water surface at a constant velocity. We will assume steady-state conditions for the convective mass transfer, constant water and air properties at 1 atm air pressure, and water surface and air temperatures which are specified later. Also, water vapor will be assumed to behave as an ideal gas in air. We will only investigate mass transfer through the water–air interface boundary layer by convective diffusion, and we will assume the water surface to be a flat plate. We will neglect all other types of mass diffusion exchange that can occur at the water–air interface due to radiation, condensation, conduction, and so on. The average rate of mass convection, M_{ave} (kg s^{-1}), from the water's surface as vapor can be defined as:

$$M_{ave} = h_{ave-mass} \times A \times (\rho_{water\,vapor\,sat} - \rho_{water\,vapor\,outside\,bl}) \qquad (19.1)$$

where $h_{ave-mass}$ (m s^{-1}) is the flat plate length averaged forced convection mass transfer coefficient between the surface of the water and the boundary layer of the flowing fluid, in this case air. A is the area of the water surface, which is assumed to be 50 m long and 20 m wide in the present case. Also, the air is flowing along the long side of the water's surface. $\rho_{water\,vapor\,sat}$ (kg m^{-3}) is the mass density of saturated water vapor on the water's surface at the water's surface temperature. $\rho_{water\,vapor\,outside\,bl}$ (kg m^{-3}) is the mass density of water vapor in air flow outside the water vapor concentration boundary layer at air temperature.

Case Studies in Fluid Mechanics with Sensitivities to Governing Variables, First Edition. M. Kemal Atesmen.
© 2019 John Wiley & Sons Ltd. This Work is a co-publication between John Wiley & Sons Ltd and ASME Press.

The forced convection mass transfer coefficient is determined by making an analogy between the momentum, heat, and mass transfers through the boundary layer. In the literature, this analogy is called the Reynolds analogy (see Ref. [15], chapters 14 and 19). The Reynolds analogy has been modified with the results of experiments, and the modified version of the Reynolds analogy for mass convection in a boundary layer over a flat plate is given as:

$$\frac{Sh_x}{Re_x \times Sc^{1/3}} = 0.5 \times C_{fx} \tag{19.2}$$

where x (m) is the distance from the leading edge of the flat plate and $Sh_x = \frac{(h_{x-mass}) \times (x)}{D_{ab}}$ is the local Sherwood number. This dimensionless number represents the forced convection mass transfer at the flat plate's surface. h_x (m s^{-1}) is the local forced convection mass transfer coefficient at the flat plate's surface and D_{ab} (m^2 s^{-1}) is the mass molecular diffusivity coefficient between the pair of species a and b, in this case water and air. $Re_x = \frac{(U_\infty) \times (x)}{v}$ is the local Reynolds number, where U_∞ (m s^{-1}) is the fluid velocity outside the boundary layer and v (m^2 s^{-1}) is the fluid's kinematic viscosity. $Sc = \frac{v}{D_{ab}}$ is the Schmidt number, which is defined as the ratio of momentum molecular diffusivity to mass molecular diffusivity, which is about 0.7 for the water–air interface. $C_{fx} = \frac{2 \times \tau_{ox}}{\rho \times U_\infty^2}$ is defined as the local skin friction coefficient, where τ_{0x} (N m^{-2}) is the local shear stress at the surface of the flat plate. The local skin friction coefficient at the surface of a flat plate for a laminar boundary layer has been determined exactly and is given in Eq. (19.3) (see Ref. [15], chapter 7):

$$C_{fx-laminar} = 0.664 \times Re_x^{-0.5} \tag{19.3}$$

By combining Eqs. (19.2) and (19.3), we can obtain the local forced convection mass transfer coefficient for a laminar flow, namely $Re_x < 500\,000$, along the surface of a flat plate as shown:

$$h_{x-mass-laminar} = 0.332 \times Sc^{1/3} \times \left(\frac{D_{ab}}{x}\right) \times Re_x^{1/2} \tag{19.4}$$

If the flow along a flat plate is totally in the laminar boundary layer region, then we can obtain a flat plate length averaged forced convection mass transfer coefficient as shown in Eqs. (19.5a) and (19.5b):

$$h_{ave-mass-laminar} = \left(\frac{1}{L}\right) \int_0^L h_{x-mass-laminar} \times dx \tag{19.5a}$$

$$h_{ave-mass-laminar} = 0.664 \times Sc^{1/3} \times \left(\frac{D_{ab}}{L}\right) \times Re_L^{1/2} \tag{19.5b}$$

The Reynolds analogy in Eq. (19.2) can be extended to the turbulent boundary layer flows along a flat plate by assuming that the molecular diffusivity and time averaged turbulent diffusivity, which is called the eddy diffusivity for mass transfer in the literature (see Ref. [9], chapter 6), are additive. The local skin friction coefficient at the surface of a flat plate for a turbulent boundary layer has been determined by experiments, and is given in Eq. (19.6) (see Ref. [9], chapter 6):

$$C_{fx-turbulent} = 0.0576 \times Re_x^{-0.2} \tag{19.6}$$

Equation (19.6) is valid for a turbulent boundary layer flow along a flat plate when $1 \times 10^5 < Re_x < 1 \times 10^7$. Combining Eqs. (19.2) and (19.6) provides the local forced convection mass transfer coefficient for a turbulent flow along the surface of a flat plate:

$$h_{x-mass-turbulent} = 0.0288 \times Sc^{1/3} \times \left(\frac{D_{ab}}{x}\right) \times Re_x^{4/5} \tag{19.7}$$

If the flow along a flat plate is totally in the turbulent boundary layer region, we can obtain a plate length averaged forced convection mass transfer coefficient as shown in Eqs. (19.8a) and (19.8b):

$$h_{ave-mass-turbulent} = \left(\frac{1}{L}\right) \int_0^L h_{x-mass-turbulent} \times dx \tag{19.8a}$$

$$h_{ave-mass-turbulent} = 0.036 \times Sc^{1/3} \times \left(\frac{D_{ab}}{L}\right) \times Re_L^{4/5} \tag{19.8b}$$

Most often, the transition from laminar boundary layer to turbulent boundary layer occurs over the flat plate. This type of transition occurs in a region, but in this analysis let us assume that the transition occurs at a point on the plate, say at length $L_{transition} < L$. In such a case, we can obtain a plate length averaged forced convection mass transfer coefficient by integrating the laminar boundary layer contribution, Eq. (19.4), from the leading edge of the plate to the transition point and adding it to the contribution from the turbulent boundary layer by integrating Eq. (19.7) from the transition point to the end of the flat plate. The plate length averaged forced convection mass transfer coefficient for this laminar plus turbulent mixed boundary layer case is shown in Eqs. (19.9a) and (19.9b):

$$h_{ave-mass-laminar \, plus \, turbulent \, bl}$$

$$= \left(\frac{1}{L}\right) \left[\int_0^{L_{transition}} h_{x-mass-laminar} \times dx + \int_{L_{transition}}^L h_{x-mass-turbulent} \times dx\right] \tag{19.9a}$$

$$h_{ave-mass-laminar \, plus \, turbulent \, bl}$$

$$= Sc^{\frac{1}{3}} \times \left(\frac{D_{ab}}{L}\right) \times \left[0.664 \times Re_{L_{transition}}^{1/2} - 0.036 \times \left(Re_{L_{transition}}^{4/5} - Re_L^{4/5}\right)\right] \tag{19.9b}$$

Now we can analyze the convection mass transfer from a water surface to the air flowing over it, where $h_{ave-mass}$ can be the flat plate length averaged forced convection mass transfer coefficient given in Eqs. (19.9a) and (19.9b) for laminar plus turbulent boundary layers. In the following analysis, we will use the following parameters and thermophysical properties:

$$L_{flat\ water\ surface\ length} = 50\,\text{m} \quad \text{and} \quad W_{flat\ water\ surface\ width} = 20\,\text{m}$$

$$T_{water\ surface} = 15°\text{C}$$

$$D_{water-air} = 2.33 \times 10^{-5}\ \text{m}^2\ \text{s}^{-1} \text{ at } 15°\text{C and 1 atm}$$

$$T_{air} = 15, 20, 25°\text{C}$$

$$v_{air\ 15°C} = 1.466 \times 10^{-5}\ \text{m}^2\ \text{s}^{-1}, v_{air\ 20°C} = 1.510 \times 10^{-5}\ \text{m}^2\ \text{s}^{-1}$$

$$v_{air\ 25°C} = 1.557 \times 10^{-5}\ \text{m}^2\ \text{s}^{-1}$$

$$\rho_{water\ vapor\ saturation\ density\ 15°C} = 0.0126\ \text{kg m}^{-3},$$

$$\rho_{water\ vapor\ saturation\ density\ 20°C} = 0.0173\ \text{kg m}^{-3}$$

$$\rho_{water\ vapor\ saturation\ density\ 25°C} = 0.0230\ \text{kg m}^{-3}$$

$$\Phi_{relative\ humidity} = 0.5$$

$\Phi_{relative\ humidity}$ is the ratio of partial pressure of water vapor in unsaturated air to partial pressure of saturated vapor at the same dry bulb temperature. We assumed that the water vapor behaves like an ideal gas in air and therefore the relative humidity can be expressed in terms of water vapor densities in air as follows:

$$\Phi_{relative\ humidity} = \frac{\rho_{water\ vapor\ outside\ bl}}{\rho_{water\ vapor\ saturated\ outside\ bl}} = 0.5 \tag{19.10}$$

Using Eqs. (19.5a), (19.5b), (19.9a), (19.9b), (19.10) and the parameters and thermophysical properties given above, we can calculate the mass convection rate of water vapor (kg h^{-1}) into air flowing in the long direction of the water's flat surface as a function of length averaged Reynolds number. The results are presented in Figure 19.1 for a constant water surface temperature of 15°C and three different dry bulb air temperatures: 15, 20, and 25°C. The rate of mass convection increases much faster in the laminar plus turbulent mixed boundary layer region than in the laminar region, as expected. The laminar plus turbulent mixed boundary layer is analyzed up to the Reynolds number of 5×10^6, which corresponds to an air velocity of $5.6\,\text{km h}^{-1}$ at 25°C. Under 50% relative humidity condition, as the flowing air dry bulb temperature rises, the mass concentration difference of water vapor across the concentration boundary layer decreases and therefore the water vapor mass convection rate decreases.

Next, let us analyze the sensitivity of water vapor mass convection rate to changes in Reynolds number. The results of these sensitivities are shown in

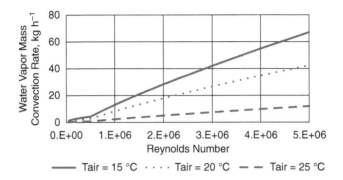

Figure 19.1   Water vapor mass convection rate versus Reynolds number for different air dry bulb temperature

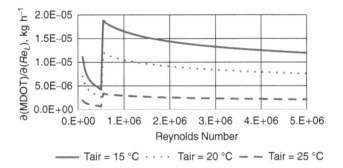

Figure 19.2   Sensitivity of water vapor mass convection rate to Reynolds number as a function of Reynolds number

Figure 19.2, again for a constant water surface temperature of 15°C and for three different dry bulb air temperatures: 15, 20, and 25°C with 50% relative humidity. The water vapor mass convection rate decreases as the Reynolds number increases, both in laminar and laminar plus turbulent flow regions. The jump in sensitivity corresponds to the transition Reynolds number from laminar to laminar plus turbulent boundary layers at 500 000. The water vapor mass convection rate sensitivity flattens out as the Reynolds number increases, and decreases very fast with increasing air dry bulb temperature.

20

Heating a Room by Natural Convection

In this chapter we will consider heating a room by natural (free) convection and by radiation using a vertical flat plate water heater which is set up perpendicular to the floor. The water heater is 2 m wide and 1 m high and it is thin. The water heater can also be set at desired uniform surface temperatures from 50 to 90°C. When the air next to the vertical water heater starts to increase in temperature, the heated air's density and therefore its body force decrease, air starts to rise, and is replaced by the higher-density, colder air in the room. This natural convection phenomenon's governing equation of motion is identical to the momentum equation for a boundary layer over a flat plate, except for a new body force term. Also, the energy equation for this natural convection phenomenon's governing equation is exactly the same as the energy equation for a boundary layer over a flat plate term (see Ref. [9], chapter 7). The new body force term in non-dimensional form represents the ratio of buoyant to viscous forces acting on a fluid element in a natural convection boundary layer, namely the Grashof number, Gr_y, (see Ref. [7]).

$$Gr_y = g \times \beta \times \left(T_s - T_{room}\right) \times Y^3/\nu^2 \qquad (20.1)$$

where g is the gravitational acceleration at the Earth's surface, $9.81\,\mathrm{m\,s^{-2}}$, $\beta\,(\mathrm{K^{-1}})$ is the coefficient of thermal expansion, defined as $1/T_{film}$ for an ideal gas, T_{film} is the average film temperature for the natural convection boundary layer, namely $0.5 \times (T_s + T_{room})$, T_s (K) is the uniform and constant surface temperature of the vertical flat plate water heater, T_{room} (K) is the averaged and uniform room

Case Studies in Fluid Mechanics with Sensitivities to Governing Variables, First Edition. M. Kemal Atesmen.
© 2019 John Wiley & Sons Ltd. This Work is a co-publication between John Wiley & Sons Ltd and ASME Press.

temperature, Y (m) is the boundary layer position on the vertical flat plate, and v (m² s⁻¹) is the kinematic viscosity of air.

For different fluids and for cases where the inertial forces can be neglected compared to the buoyancy forces, a product of the Grashof number and the Prandtl number Pr, called the Rayleigh number Ra_y, is used for experimental correlations. $Pr = \frac{v}{\alpha}$ is the Prandtl number defined as the ratio of momentum molecular diffusivity to thermal molecular diffusivity, where $\alpha = \frac{k}{\rho \times c_p}$, k [W (m K)⁻¹] is the thermal conductivity of air, ρ (kg m⁻³) is the density of air, and c_p [(W s) (kg K)⁻¹] is the specific heat of air at constant pressure. In the present analysis, all thermophysical properties of air are considered at T_{film} temperature.

In the present analysis, it is also assumed that the transition from the laminar natural convection boundary layer to the turbulent natural convection boundary layer occurs abruptly at $Ra_y = Pr \times Gr_x = 1 \times 10^9$. For a vertical plate water heater's local natural convection laminar heat transfer coefficient between two vertical surfaces (front and back) and the surrounding room air, the following analytically driven equation is used:

$$h_{y-laminar} = 0.360 \times \left(\frac{k}{Y}\right) \times Gr_y^{0.25} \tag{20.2}$$

For a vertical plate water heater's local natural convection turbulent heat transfer coefficient between two vertical surfaces (front and back) and the surrounding room air, the following experimentally determined equation is used:

$$h_{y-turbulent} = 0.0218 \times \left(\frac{k}{Y}\right) \times Gr_y^{0.40} \tag{20.3}$$

For a vertical plate water heater with height H, the averaged natural convection laminar heat transfer coefficient is obtained by integrating Eq. ((20.2)) to give:

$$h_{average-laminar} = \left(\frac{1}{H}\right) \int_0^H h_{y-laminar} dy = 0.48 \times \left(\frac{k}{H}\right) \times Gr_H \tag{20.4}$$

In the present vertical plate water heater, the natural convection boundary layer is mixed. It starts as a laminar boundary layer at the bottom of the plate and becomes turbulent at a height of 0.36 m for a surface temperature of 90°C with the assumption that the transition occurs abruptly at $Ra_y = Pr \times Gr_x = 1 \times 10^9$. The vertical plate height averaged natural convection heat transfer coefficient for this laminar plus turbulent mixed boundary layer case is:

$$h_{ave-mixed\ bl} = \left(\frac{1}{H}\right) \left[\int_0^{H_{transition}} h_{y-laminar} dy + \int_{H_{transition}}^H h_{y-turbulent} dy \right]$$

$$h_{ave-mixed\ bl} = \left(\frac{k}{H}\right) \left[0.48 Gr_{H_{transition}}^{0.25} - 0.0182 \left(Gr_{H_{transition}}^{0.4} - Gr_H^{0.4} \right) \right] \tag{20.5}$$

Let us now evaluate the average room air temperature increase in the room due to natural convection heat transfer and radiation heat transfer from the vertical plate

water heater to the room air, and the heat lost from the room air to the outside
environment. The following energy balance can be constructed to highlight the
transient behavior of the average room air temperature:

$$\rho c_p V \frac{dT_{room}}{dt} = h_{ave-mixed\ bl} A_s (T_s - T_{room}) + \varepsilon \sigma A_s (T_s^4 - T_{room}^4)$$

$$- h_{ave-room\ to\ environment} A_{room} (T_{room} - T_{environment}) \qquad (20.6)$$

where $V = 75$ m^3 is the volume of air that has to be heated in the room (assuming
no air leaks from the room), T_{room} (°C) is the instantaneous room temperature at
a given time, initially 10°C, t (s) is the time that the vertical plate water heater has
been heating the room air, $T_s = 50, 60, 70, 80, 90$°C are the vertical plate water
heater's constant surface temperatures used in the present calculations, $A_s = 4$ m^2
is the vertical plate water heater's surface area at constant temperature, $\varepsilon = 0.9$ is
the surface emissivity of the vertical plate water heater for surface temperatures
between 50 and 90°C, $\sigma = 5.67 \times 10^{-8}$ W (m^2 K^4)$^{-1}$ is the Stefan–Boltzmann
constant for blackbody radiation, $h_{ave-room\ to\ environment} = 2$ W (m^2 K)$^{-1}$ is the
average heat loss coefficient between the room air and the outside environment,
$A_{room} = 60$ m^2 is the total area of the room's walls, floor, and ceiling for heat loss
to the environment, and $T_{environment} = 0$°C.

The room temperature versus time is obtained by solving Eq. (20.6) by numer-
ical integration. All thermophysical properties for room air are considered at the
instantaneous film temperature, namely $0.5 \times (T_s + T_{room})$ at time t. The room temperature versus time for five different vertical plate water heater surface temperatures are shown in Figure 20.1. As the heater surface temperature increases, the
room air temperature increases very fast. Hopefully, the heater thermostat will
shut the heater off at the desired comfortable room temperature. Figure 20.1 shows
that, as time advances, the room temperature increase slows down. This slowing

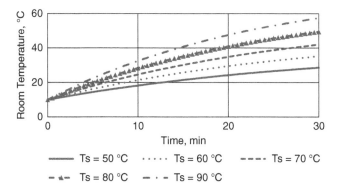

Figure 20.1 Room temperature versus time for five different vertical plate water heater
surface temperatures

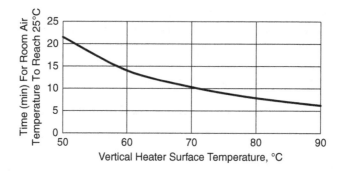

Figure 20.2 Time (min) for room air temperature to reach 25°C versus vertical heater surface temperature

down is due to a decrease in both natural convection heat transfer and radiation heat transfer. As the room temperature increases, the buoyant forces acting on the air particles decrease. With decreasing $(T_s - T_{room})$ and coefficient of thermal expansion, the instantaneous Grashof number decreases and therefore the laminar plus turbulent natural convection heat transfer coefficient decreases. With decreasing $(T_s^4 - T_{room}^4)$, the radiation heat transfer decreases. With advancing time, the only heat transfer increase is from room air to the environment, namely due to the increasing heat transfer potential $(T_{room} - T_{environment})$.

The time for the room air temperature to reach the desired 25°C versus vertical heater surface temperature is presented in Figure 20.2. When the vertical heater surface temperature is 50°C, it takes 21.5 min to heat the room air to the desired temperature. When the vertical heater surface temperature is 90°C, it takes 6.2 min to heat the room air to the desired temperature.

The average natural convection heat transfer coefficient for laminar plus turbulent boundary layers versus room air temperature for $T_s = 90$°C is shown in Figure 20.3. With decreasing $(T_s - T_{room})$, the Grashof number decreases and therefore the average natural convection heat transfer coefficient decreases.

Different heat transfer modes versus time for a vertical heater temperature of 90°C are shown in Figure 20.4. As time progresses (i.e. around 30 min), the natural convection heat transfer decreases, and the room air is heated only by radiative heat transfer. Natural convection heat transfer from the vertical heater surface compensates the heat loss to the environment. Different heat transfer modes' percentage of Q_{net} versus time for the vertical heater surface temperature of 90°C are shown in Figure 20.5. The percentage of natural convection heat transfer from the vertical heater surface to Q_{net} stays constant for the first 30 min of the heating process for a vertical heater surface temperature of 90°C.

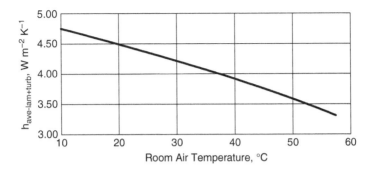

Figure 20.3 Average natural convection heat transfer coefficient for laminar plus turbulent boundary layers versus room air temperature for $T_s = 90°C$

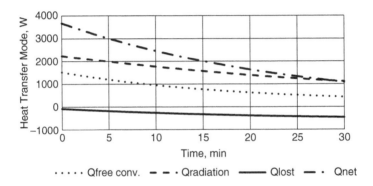

Figure 20.4 Different heat transfer modes versus time for vertical heater surface temperature of 90°C

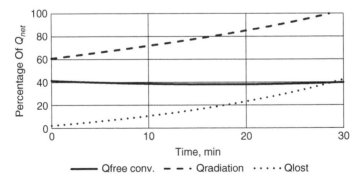

Figure 20.5 Different heat transfer modes' percentage of Q_{net} versus time for vertical heater surface temperature of 90°C

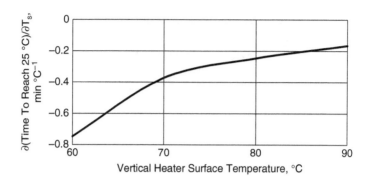

Figure 20.6 Sensitivity of time to reach 25°C room air temperature to vertical heater surface temperature as a function of vertical heater surface temperature

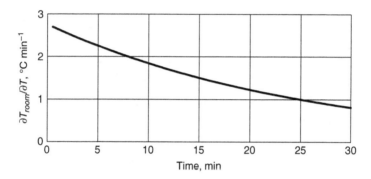

Figure 20.7 Sensitivity of room air temperature to heating time as a function of heating time for $T_s = 90°C$

The sensitivity of changes in time to reach 25°C room air temperature to changes in vertical heater surface temperature as a function of vertical heater surface temperature is shown in Figure 20.6. As the vertical heater surface temperature increases, the absolute value of the time sensitivity to reach 25°C decreases.

The sensitivity of changes in room air temperature to changes in heating time as a function of heating time for $T_s = 90°C$ is shown in Figure 20.7. This sensitivity decreases as time advances. Initially this sensitivity was 2.7°C min^{-1}; after 30 min it decreased to 0.8°C min^{-1}.

The sensitivity of changes in average laminar plus turbulent boundary layer natural convection heat transfer coefficient to changes in temperature difference

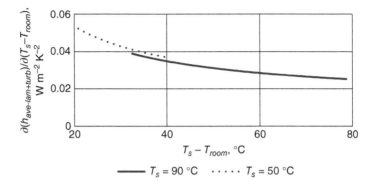

Figure 20.8 Sensitivity of average laminar plus turbulent boundary layer natural convection heat transfer coefficient to temperature difference between the vertical heater surface and room air

between the vertical heater surface and room air is presented in Figure 20.8 for two different vertical heater surface temperatures. As the room temperature increases and therefore $(T_s - T_{room})$ decreases, the reduction in natural convection heat transfer coefficient increases. This sensitivity is more prominent for lower vertical heater plate surface temperatures.

21

Laminar Flow Through Porous Material

In this chapter we will consider laminar flow through porous material beds. Throughout history, flows through porous media have been studied both experimentally and theoretically. The dependent variables that have to be determined are the pressure drop ΔP (N m^{-2}) through the porous material bed and the fluid volume flow rate Q (m^3 s^{-1}) as functions of porous material void fraction, characteristic void diameter, porous material bed length and area, and fluid viscosity. Henri Darcy [3] performed sand column experiments in the nineteenth century in order to determine the ΔP versus Q relationship for sand columns. Further investigations by Carmen [2] and Kozeny [8] related the ΔP versus Q relationship to fluid viscosity and porous material bed characteristics for laminar flows.

The total volume flow rate through the porous material bed can be related to the total flow rate through the voids by the following continuity relationship:

$$Q = \overline{U}_{bed} \times A = \overline{U}_{void} \times A_{void} \qquad (21.1)$$

where $\overline{U}_{bed}$ (m s^{-1}) is the superficial average fluid velocity coming out from the bottom of the porous material bed, A (m^2) is the bottom cross-sectional area of the porous material bed, $\overline{U}_{void}$ (m s^{-1}) is the average fluid velocity in voids, and A_{void} (m^2) is the total average cross-sectional area of voids through the porous material bed. Let us define the fraction of voids ε in the porous material bed as:

$$\varepsilon = \frac{A_{void} \times L_{void}}{A \times L} = \frac{A_{void}}{A} = \frac{\overline{U}_{bed}}{\overline{U}_{void}} \qquad (21.2)$$

Case Studies in Fluid Mechanics with Sensitivities to Governing Variables, First Edition. M. Kemal Atesmen.

In Eq. (21.2) it is assumed that the average length of voids from the top of the bed to the bottom of the bed is equal to the length L (m) from the top of the porous material bed to the bottom of the porous material bed. For a cubic porous material bed that is packed with the same diameter (d) spherical particles, the fraction of voids is given by:

$$\varepsilon = \frac{d^3 - \left(\frac{\pi}{6}\right) \times d^3}{d^3} = 0.4764 \tag{21.3}$$

Since the void diameters change throughout the porous material bed, we have to define a characteristic void diameter d_{void} (m), as shown:

$$d_{void} = \frac{A \times \varepsilon \times L}{A \times (1 - \varepsilon) \times L \times S} \tag{21.4}$$

where S (m^{-1}) is the specific surface area of voids per unit volume. For a cubic porous material bed that is packed with the same diameter (d) spherical particles, S is given by:

$$S = \frac{\pi \times d^2}{\left(\frac{\pi}{6}\right) \times d^3} = \frac{6}{d} \tag{21.5}$$

Let us assume that the flows through voids behave like a fully developed laminar flow in a pipe, see Chapter 16. Then the pressure drop through the void pipes can be written as:

$$\Delta P = \frac{32 \times L \times \mu \times \overline{U}_{void}}{d_{void}^2} = \frac{K \times L \times \mu \times \overline{U}_{void}}{d_{void}^2} \tag{21.6}$$

where μ [(N s) m^{-2}] is the viscosity of the fluid. Combining Eqs. (21.2), (21.4), and (21.6) and replacing the constant by the experimentally obtained Kozeny constant, $K = 5$, Eq. (21.7) can be obtained for laminar flow through a porous material bed packed with the same diameter (d) spherical particles:

$$\frac{\Delta P}{Q} = 180 \times \left(\frac{\mu \times L}{d^2 \times A}\right) \times \left[\frac{(1 - \varepsilon)^2}{\varepsilon^3}\right] \tag{21.7}$$

$\frac{\Delta P}{Q}$ given in Eq. (21.7) is valid in laminar flow through porous media beds. The flow condition, namely the Reynolds number for laminar flow through a void passage, has to be less than 2, as shown in Eq. (21.8):

$$Re_d = \frac{\overline{U}_{void} \times d_{void}}{v} = \frac{Q \times d}{6 \times A \times (1 - \varepsilon) \times v} < 2 \tag{21.8}$$

where v (m^2 s^{-1}) is the kinematic viscosity of the fluid. Let us investigate Eq. (21.7) by using four different fluids flowing through voids in porous material beds with different void fractions. These four fluids are presented in Table 21.1.

Table 21.1   Viscosities of four fluids used in the present analysis

Fluids at 20°C and 1 atm	Viscosity μ [(N s) m^{-2}]	Kinematic viscosity v (m^2 s^{-1})
Water	0.001	1.00×10^{-6}
Gasoline	0.00029	4.03×10^{-7}
Motor oil	0.03	3.26×10^{-5}
Glycerin	1.5	1.19×10^{-3}

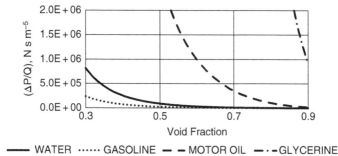

Figure 21.1 $\frac{\Delta P}{Q}$ for different fluids versus void fraction

$\frac{\Delta P}{Q}$ versus void fraction is investigated using the four fluids given in Table 21.1 along with the following parameters:

$$L = 1\,\text{m}, A = 4\,\text{m}^2, d = 0.001\,\text{m}$$

In all investigations carried out in this chapter, we insure that all calculations are in the laminar flow regime, namely Eq. (21.8) is satisfied. The results obtained for pressure drop-to-fluid volume flow ratio versus void fraction values between 0.3 and 0.9 are shown in Figures 21.1 and 21.2. As the viscosity of the fluid increases, the pressure drop-to-fluid volume flow ratio becomes more sensitive to changes in void fraction. The effects of viscosity on pressure drop-to-fluid volume flow ratio are more prominent at low values of void fraction. An increase in void fraction of the porous material bed causes a decrease in pressure drop-to-fluid volume flow rate ratio, as shown in Figures 21.3 and 21.4. The sensitivity of pressure drop-to-fluid volume flow rate ratio becomes significant at low values of void fraction. A positive change in void fraction causes the absolute value of $\frac{\Delta P}{Q}$ to decrease in proportion to $\left[-\frac{(\varepsilon-3)(\varepsilon-1)}{\varepsilon^4} \right]$.

$\frac{\Delta P}{Q}$ versus porous material bed length is investigated using the four fluids given in Table 21.1 along with the following parameters:

$$\varepsilon = 0.4, A = 4\,\text{m}^2, d = 0.001\,\text{m}$$

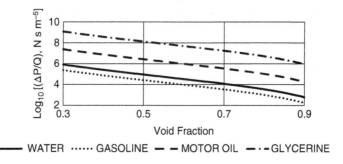

Figure 21.2 $Log_{10} \left[\frac{\Delta P}{Q} \right]$ for different fluids versus void fraction

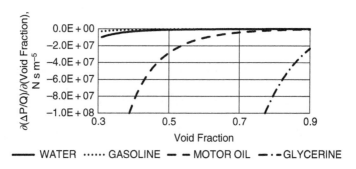

Figure 21.3   Sensitivity of $\frac{\Delta P}{Q}$ to void fraction as a function of void fraction

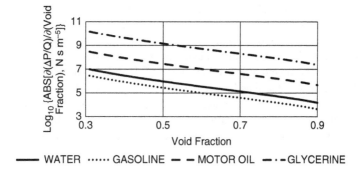

Figure 21.4 $Log_{10} \left[\text{Absolute value of sensitivity of } \frac{\Delta P}{Q} \text{ to void fraction} \right]$ as a function of void fraction

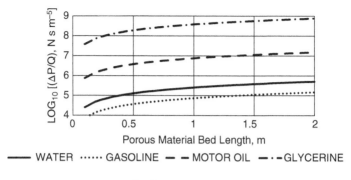

Figure 21.5 $\text{Log}_{10}\left[\frac{\Delta P}{Q}\right]$ versus porous material bed length

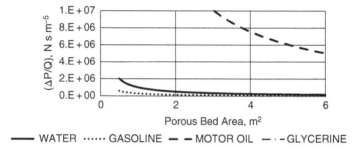

Figure 21.6 $\frac{\Delta P}{Q}$ for different fluids versus porous material bed area. Glycerin's $\frac{\Delta P}{Q}$ is above the range of this plot

The results obtained for $\log_{10}$ of the pressure drop-to-fluid volume flow ratio versus porous material bed length values between 0.1 and 2.0 m are shown in Figure 21.5. For a given fluid, the increase in pressure drop-to-fluid volume flow ratio over the increase in the porous material bed length is a positive constant. This positive constant slope is 73 406 (N s) m^{-6} for gasoline, 253 125 (N s) m^{-6} for water, 7 593 750 (N s) m^{-6} for motor oil, and 379 687 500 (N s) m^{-6} for glycerin.

$\frac{\Delta P}{Q}$ versus porous material bed area is investigated using the four fluids given in Table 21.1 along with the following parameters:

$$\varepsilon = 0.4, L = 1 \text{ m}, d = 0.001 \text{ m}$$

The results obtained for pressure drop-to-fluid volume flow ratio versus porous material bed area for values between 0.5 and 6.0 m^2 are shown in Figures 21.6 and 21.7. For a given fluid, the pressure drop-to-fluid volume flow rate decreases proportionally to $\frac{1}{A}$ as the porous material bed area increases.

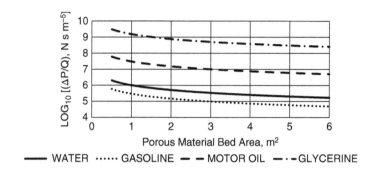

Figure 21.7 $\text{Log}_{10}\left[\frac{\Delta P}{Q}\right]$ for different fluids versus porous material bed media area

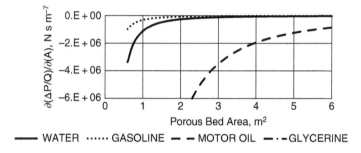

Figure 21.8 Sensitivity of $\frac{\Delta P}{Q}$ to porous material bed area versus porous material bed area. The sensitivity of glycerin is beyond the range of this plot

Changes in $\frac{\Delta P}{Q}$ with respect to changes in porous material bed area are presented in Figures 21.8 and 21.9. The sensitivity of changes in $\frac{\Delta P}{Q}$ with respect to changes in the porous material bed area behaves like $\left(-\frac{1}{A^2}\right)$. As the porous material bed area increases, the absolute value of the change in $\frac{\Delta P}{Q}$ decreases. For higher-viscosity fluids, the $\frac{\Delta P}{Q}$ values are more sensitive to changes in porous material bed area.

$\frac{\Delta P}{Q}$ versus characteristic spherical void diameter is investigated using the four fluids given in Table 21.1 along with the following parameters:

$$A = 4\,\text{m}^2, L = 1\,\text{m}, \varepsilon = 0.4$$

The results obtained for pressure drop-to-fluid volume flow ratio versus characteristic spherical void diameter for values between 0.0001 and 0.005 m are shown in Figure 21.10. For a given fluid, the pressure drop-to-fluid volume flow

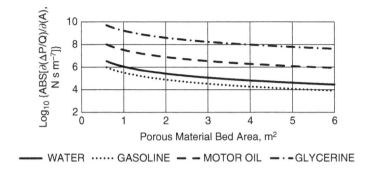

Figure 21.9 Log_{10} $\left[$Absolute value of sensitivity of $\frac{\Delta P}{Q}$ to porous material bed area$\right]$ versus porous material bed area

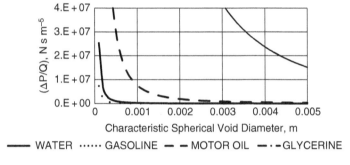

Figure 21.10 $\frac{\Delta P}{Q}$ versus characteristic spherical void diameter

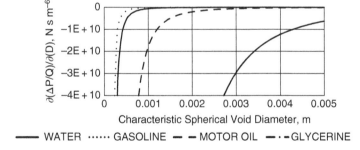

Figure 21.11 Sensitivity of $\frac{\Delta P}{Q}$ to characteristic spherical void diameter versus characteristic spherical void diameter

rate decreases proportionally to $\frac{1}{d^2}$ as the characteristic spherical void diameter increases. As expected, the pressure drop-to-fluid volume flow rate decrease is more pronounced at small characteristic spherical void diameters.

The changes in $\frac{\Delta P}{Q}$ with respect to changes in characteristic spherical void diameter are presented in Figure 21.11. The sensitivity of the changes in $\frac{\Delta P}{Q}$ with respect to the changes in characteristic spherical void diameter behaves like $\left(-\frac{1}{d^3}\right)$. As the characteristic spherical void diameter increases, the absolute value of the change in $\frac{\Delta P}{Q}$ decreases. For higher-viscosity fluid, the $\frac{\Delta P}{Q}$ values are more sensitive to changes in characteristic spherical void diameter.

22

Condensation on the Surface of a Vertical Plate in a Laminar Flow Regime

In this chapter we will consider laminar flow of the condensate film of water vapor on the surface of a vertical plate. Condensation occurs if the water vapor is cooled below its saturation temperature. During condensation, latent heat of condensation is released and is assumed to be transferred from the water vapor–liquid condensate film interface to the cooler vertical plate surface only by conduction. The condensate film starts at zero thickness at the top of the vertical plate and starts to flow downwards under the action of gravity while the condensate film thickness grows. In the present analysis, we will assume a laminar flow of the condensate film, namely the Reynolds number will be less than 2000, and we will also assume that the condensate film is flowing downwards smoothly without any waviness. The heat transfer coefficient for a condensate film in a laminar flow was first developed by Nusselt in 1916 (see Ref. [9], chapter 10).

The local conduction heat transfer from the water vapor–liquid film interface to the cooler vertical plate is defined in Eq. (22.1). This shows that the local temperature gradient through the thickness of the condensate film is linear. It is assumed that the vertical plate's surface temperature T_{plate} is a constant and is less than the water vapor saturation temperature $T_{saturation}$, which is 100°C at 1 atm:

$$q(z) = \frac{k_l}{\delta(z)}(T_{saturation} - T_{plate}) \qquad (22.1)$$

Case Studies in Fluid Mechanics with Sensitivities to Governing Variables, First Edition. M. Kemal Atesmen.
© 2019 John Wiley & Sons Ltd. This Work is a co-publication between John Wiley & Sons Ltd and ASME Press.

where $q(z)$ $(W\,m^{-2})$ is the local heat transfer, z (m) is the downward distance from the top edge of the vertical plate, k_l $[W\,(m\,K)^{-1}]$ is the thermal conductivity of the condensate film at the average film temperature, namely at $0.5 \times (T_{saturation} - T_{plate})$, and $\delta(z)$ (m) is the local condensate film thickness, which is zero at $z = 0$. The local heat transfer coefficient $h(z)$ is defined as follows:

$$q(z) = h(z) \times (T_{saturation} - T_{plate}) \tag{22.2}$$

Combining Eqs. (22.1) and (22.2) shows that the local heat transfer coefficient is only a function of the local condensate film thickness, while assuming a constant thermal conductivity of the condensate film as shown:

$$h(z) = \frac{k_l}{\delta(z)} \tag{22.3}$$

The velocity profile for the condensate film can be obtained by a force balance on an element of the condensate film by equating the shear stress with the forces of gravity and buoyancy for an incompressible and constant property laminar flow while neglecting advection (see Ref. [9], chapter 10). The local condensate film velocity profile turns out to be parabolic, as shown in Eq. (22.4):

$$U(y) = \left(\frac{g}{\mu_l}\right) \times (\rho_l - \rho_v) \times [\delta(z) \times y - y^2] \tag{22.4}$$

where y is the perpendicular distance from the vertical plate's surface through the condensate film and the velocity is zero at $y = 0$. $g = 9.81$ m s^{-2} is the gravitational constant at sea level, μ_l $[(N\,s)\,m^{-2}]$ is the viscosity of the condensate film at the average film temperature, ρ_l $(kg\,m^{-3})$ is the density of the condensate film at the average film temperature, ρ_v $(kg\,m^{-3})$ is the density of water vapor at the saturation vapor temperature. The condensate film mass flow rate (kg s^{-1}) through a horizontal cross-section of the vertical plate with width W is given by:

$$\dot{m}(z) = \rho_l \times \overline{U}(z) \times \delta(z) \times W \tag{22.5a}$$

where the mean condensate film velocity is:

$$\overline{U}(z) = \frac{1}{\delta(z)} \times \int_0^{\delta(z)} U(y) \times dy. \tag{22.5b}$$

Integration of Eq. (22.5a) gives the local condensate film mass flow rate:

$$\dot{m}(z) = \frac{g \times W}{3 \times \mu_l} \times \rho_l \times (\rho_l - \rho_v) \times \delta^3(z) \tag{22.6}$$

An incremental increase in condensate film mass flow rate requires an incremental transfer of latent heat of condensation h_{fg} (J kg^{-1}) to the vertical wall surface, as shown in Eqs. (22.7a) and (22.7b). In the present example, $h_{fg} = 2\,257\,000$ J kg^{-1} for water vapor at $T_{saturation} = 100°C$ and 1 atm:

$$h_{fg} \times d\dot{m}(z) = q(z) \times W \times dz \tag{22.7a}$$

$$h_{fg} \times \frac{d\dot{m}(z)}{dz} = \frac{k_l \times W}{\delta(z)} \times (T_{saturation} - T_{plate}) \tag{22.7b}$$

Combining Eqs. (22.6), (22.7a), and (22.7b) and integrating between $\delta(z) = 0$ at $z = 0$ and $\delta(z) = \delta$ at $z = z$, the following condensate film thickness is obtained as a function of distance z from the top edge of the vertical plate:

$$\delta^4(z) = \frac{[4 \times k_l \times \mu_l \times (T_{saturation} - T_{plate}) \times z]}{[g \times h_{fg} \times \rho_l \times (\rho_l - \rho_v)]} \tag{22.8}$$

Using Eqs. (22.3) and (22.8), the local heat transfer coefficient can be obtained as:

$$h^4(z) = \frac{[g \times h_{fg} \times k_l^3 \times \rho_l \times (\rho_l - \rho_v)]}{[4 \times \mu_l \times (T_{saturation} - T_{plate}) \times z]} \tag{22.9}$$

The local heat transfer coefficient given in Eq. (22.9) can be integrated from the top of the vertical plate to the bottom through a plate height of H in order to obtain the average heat transfer coefficient for the whole vertical plate:

$$h_{average} = \frac{1}{H} \int_0^H h(z) \times dz = 0.943 \times \left[\frac{g \times h_{fg} \times k_l^3 \times \rho_l \times (\rho_l - \rho_v)}{\mu_l \times (T_{saturation} - T_{plate}) \times H} \right]^{0.25} \tag{22.10}$$

In the present analysis, we are analyzing the laminar flow regime of the condensate film over the vertical plate, namely the maximum Reynolds number has to be less than 2000. The maximum Reynolds number occurs at the bottom of the vertical plate at $z = H$, where the condensate film thickness and therefore the mass flow rate are maximum:

$$Re_H = \frac{\rho_l \times U_{mean\ at\ H} \times D_H}{\mu_l} = \frac{4 \times g \times \rho_l \times (\rho_l - \rho_v) \times \delta_H^3}{3 \times \mu_l^2} \tag{22.11}$$

The following example for laminar flow of a condensate film of water vapor on the surface of a vertical plate is performed for a plate with $H = 2$ m, $W = 2$ m, and all condensate liquid film properties taken at the average film temperature, namely $0.5 \times (T_{saturation} - T_{plate})$. The vertical plate height versus condensate film thickness for different vertical plate constant temperatures is obtained using Eq. (22.8), and the results are presented in Figure 22.1. As the vertical plate temperature decreases the condensate film thickness (μm) increases, as expected.

Figure 22.2 shows the sensitivities of condensate film thickness to vertical distance from the top edge of the vertical plate. These sensitivities are high at the top

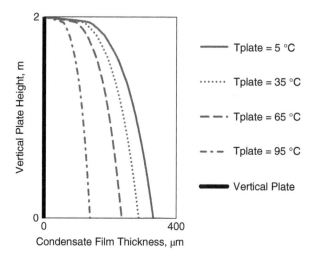

Figure 22.1 Vertical plate height versus condensate film thickness for different plate temperatures

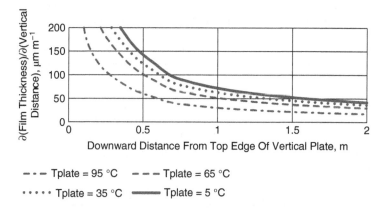

Figure 22.2 Sensitivity of film thickness to vertical distance as a function of downward distance z from the top edge of the vertical plate

of the vertical plate and decrease as $z^{-0.75}$ as the condensate film moves down the plate. These sensitivities also increase as $(T_{saturation} - T_{plate})$ increases.

Since the present analysis stays within the laminar flow regime, it is a good idea to check the maximum Reynolds number using Eq. (22.11). The maximum Reynolds number occurs at the bottom of the vertical plate when the plate temperature is 25°C (i.e. $Re_H = 1714$). As the vertical plate surface temperature increases,

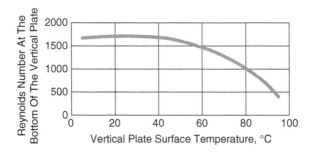

Figure 22.3   Maximum Reynolds number versus vertical plate surface temperature

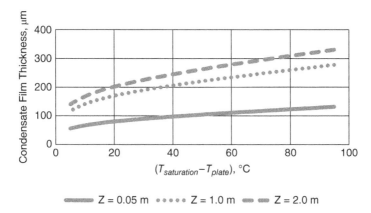

Figure 22.4   Condensate film thickness versus $(T_{saturation} - T_{plate})$ at three different vertical plate locations

δ_H decreases and so does the Reynolds number, even if the condensate liquid film properties' ratio is increasing, namely $\left[\dfrac{\rho_l \times (\rho_l - \rho_v)}{\mu_l^2}\right]$ is increasing (see Figure 22.3).

The condensate film thickness versus $(T_{saturation} - T_{plate})$ at three different vertical plate locations is shown in Figure 22.4. These curves are obtained from Eq. (22.8). As $(T_{saturation} - T_{plate})$ increases, so does the condensate film thickness. The sensitivity of the condensate film thickness to $(T_{saturation} - T_{plate})$ decreases as $(T_{saturation} - T_{plate})$ increases, as shown in Figure 22.5. The condensate film thickness to $(T_{saturation} - T_{plate})$ decrease rate is $(T_{saturation} - T_{plate})^{-0.75}$.

The condensate mass flow rate versus $(T_{saturation} - T_{plate})$ for the whole vertical plate is obtained using Eqs. (22.7a), (22.7b), and (22.10) and the results are shown in Figure 22.6. The condensate mass flow rate increases as $(T_{saturation} - T_{plate})$ increases, as expected. The sensitivity of the condensate mass flow rate to

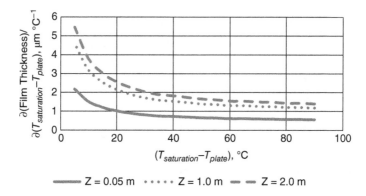

Figure 22.5 Sensitivity of condensate film thickness to $(T_{saturation} - T_{plate})$ at three different vertical plate locations

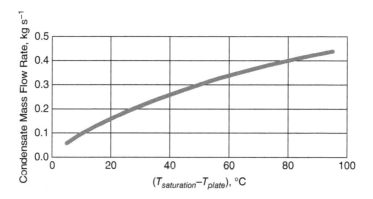

Figure 22.6 Condensate mass flow rate versus $(T_{saturation} - T_{plate})$

$(T_{saturation} - T_{plate})$ is presented in Figure 22.7. This sensitivity is high at low $(T_{saturation} - T_{plate})$ values and decreases as $(T_{saturation} - T_{plate})$ increases at a rate of $(T_{saturation} - T_{plate})^{-0.25}$.

The average heat transfer coefficient decreases as $(T_{saturation} - T_{plate})$ increases. The results obtained using Eq. (22.10) are shown in Figure 22.8. The sensitivity of the average heat transfer coefficient to $(T_{saturation} - T_{plate})$ is presented in Figure 22.9. The absolute value of this sensitivity decreases as $(T_{saturation} - T_{plate})$ increases, namely

$$\frac{\partial h_{average}}{\partial (T_{saturation} - T_{plate})} \propto (T_{saturation} - T_{plate})^{-1.25}$$

The local heat transfer coefficient decreases as the downward distance z from the top of the vertical plate increases. The results obtained using Eq. (22.9) are

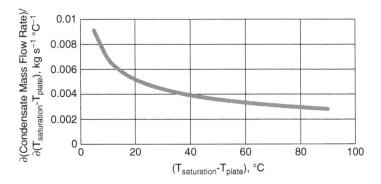

Figure 22.7   Sensitivity of condensate mass flow rate to $(T_{saturation} - T_{plate})$

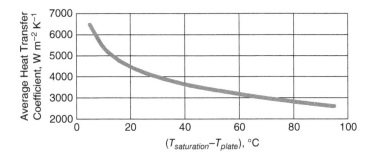

Figure 22.8   Average heat transfer coefficient versus $(T_{saturation} - T_{plate})$

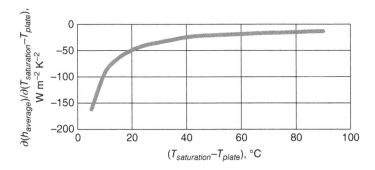

Figure 22.9   Sensitivity of average heat transfer coefficient to $(T_{saturation} - T_{plate})$

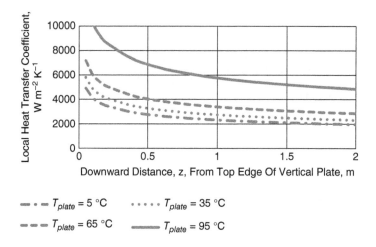

Figure 22.10 Local heat transfer coefficient versus downward distance z from the top edge of the vertical plate

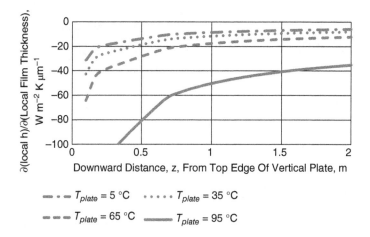

Figure 22.11 Sensitivity of local heat transfer coefficient to local condensate film thickness

shown in Figure 22.10 for four different plate temperatures. The sensitivity of the local heat transfer coefficient to local condensate film thickness is presented in Figure 22.11. The absolute value of this sensitivity decreases as the downward distance z from the top of the vertical plate increases, namely $\frac{\partial h(z)}{\partial \delta(z)} = -k_l \times \delta(z)^{-2}$. Also, the absolute value of this sensitivity increases as $(T_{saturation} - T_{plate})$ decreases.

23

A Non-Newtonian Fluid Flow in a Pipe

In this chapter we will consider time-independent non-Newtonian fluids flowing in a pipe in a fully developed state and in a laminar flow regime. Non-Newtonian fluids are studied in rheology, which is the science of deformation and fluid flow. In Newtonian fluids like water, air, milk, and so on, the shear stress applied to a fluid element is proportional to the shear rate, namely the local velocity gradients, where the proportionality constant is the fluid's viscosity μ [(N s) m^{-2}], constant at a given temperature. On the other hand, for non-Newtonian fluids, the fluid's viscosity is not a constant at a given temperature and changes with the magnitude of the shear rate and in some cases with time. Most non-Newtonian fluids follow the power law relationship between the shear stress and the shear rate as given in Eq. (23.1), which is known as the Ostwald–de Waele equation (see Ref. [14], chapter 9):

$$\tau = K \times \left(-\frac{dU}{dr} \right)^n \tag{23.1}$$

Equation (23.1) changes along the radial direction of the pipe and also with the longitudinal flow direction in the pipe. τ (N m^{-2}) is the shear stress exerted on the fluid element, K [(N s^n) m^{-2}] is the flow consistency coefficient, n is the flow behavior index, and $\frac{dU}{dr}$ (s^{-1}) is the velocity gradient in the radial direction, commonly called the shear rate. Equation (23.1) can also be rewritten by introducing an apparent viscosity $\mu_{apparent}$ [(N s) m^{-2}]:

$$\tau = \mu_{apparent} \times \left(-\frac{dU}{dr} \right) \tag{23.2}$$

Case Studies in Fluid Mechanics with Sensitivities to Governing Variables, First Edition. M. Kemal Atesmen.
© 2019 John Wiley & Sons Ltd. This Work is a co-publication between John Wiley & Sons Ltd and ASME Press.

where $\mu_{apparent} = K \times \left(\frac{dU}{dr}\right)^{n-1}$. When $n < 1$, non-Newtonian fluids are categorized as pseudoplastic or shear thinning fluids such as carrot puree, yoghurt, mayonnaise, and toothpaste. Most non-Newtonian fluids exhibit pseudoplastic behavior. When $n > 1$, non-Newtonian fluids are categorized as dilatant fluids or shear thickening fluids such as wet cement and quicksand. When $n = 1$, fluids behave as Newtonian fluids and $\mu_{apparent} = K = \mu$.

There is another type of non-Newtonian fluid, like paint, slurry, or ketchup, which requires an initial shear stress τ_0 to be applied before the shear rate starts to increase and the shear stress-to-shear rate ratio behaves as a constant. These types of non-Newtonian fluids are called Bingham plastics. The shear stress versus shear rate behavior of different types of non-Newtonian fluids and Newtonian fluids is shown in Figure 23.1.

Figure 23.1 shows that $\mu_{apparent}$ decreases as the shear rate increases for pseudoplastic fluids, whereas $\mu_{apparent}$ increases as the shear rate increases for dilatant fluids. For Newtonian fluids and Bingham plastics, the shear stress divided by the shear rate shows constant behavior.

For a fully developed laminar flow in a pipe, a force balance on a cylindrical fluid element with radius r and length L, centered in the middle of the pipe, can be performed between the pressure drop and the shear stress, namely $(2 \times \pi \times r \times L \times \tau) + (\pi \times r^2 \times \Delta P) = 0$. In this force balance, ΔP ($\mathrm{N\,m^{-2}}$) is the pressure drop which accelerates the fluid element against the frictional shear stress. This force balance provides us with a first-order differential equation, as shown in Eq. (23.3), between the shear rate and the radial direction in the pipe for non-Newtonian fluids that follow the power law given by Eq. (23.1):

$$\frac{dU}{dr} = -\left(-\frac{\Delta P}{2 \times K \times L}\right)^{1/n} \times r^{1/n} \qquad (23.3)$$

Integrating Eq. (23.3) and using the boundary condition $U = 0$ at $r = R$ provides us with the velocity profile for non-Newtonian fluids that follow the power law

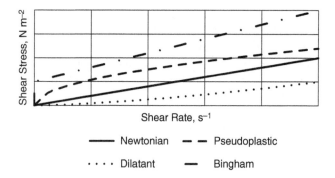

Figure 23.1   Shear stress versus shear rate for different types of fluids

given by Eq. (23.1) for a fully developed laminar flow in a pipe. This velocity profile is presented as:

$$U = \left(-\frac{\Delta P}{2 \times K \times L}\right) \times \left(\frac{n}{n+1}\right) \times R^{\frac{n+1}{n}} \times \left[1 - \left(\frac{r}{R}\right)^{\frac{n+1}{n}}\right] \quad (23.4)$$

Equation (23.4) reduces to a parabolic velocity profile over the pipe's radius for $n = 1$. The mean velocity over the cross-section of the pipe is provided in Eq. (23.5). The mean velocity is obtained by integrating Eq. (23.4) over the cross-sectional area of the pipe and normalizing the result by dividing the result by $\pi \times R^2$:

$$\overline{U} = \left(-\frac{\Delta P}{2 \times K \times L}\right)^{\frac{1}{n}} \times \left(\frac{n}{3 \times n + 1}\right) \times R^{\frac{n+1}{n}} \quad (23.5)$$

The velocity profile given in Eq. (23.4) can be rewritten by using the above mean velocity expression:

$$U = \overline{U} \times \left(\frac{3 \times n + 1}{n + 1}\right) \times \left[1 - \left(\frac{r}{R}\right)^{\frac{n+1}{n}}\right] \quad (23.6)$$

Several fluid velocity profiles in a fully developed laminar pipe flow are presented in Figure 23.2. The velocity profiles are shown from the center of the pipe at $r = 0$ to its wall at $r = R$, and they are symmetric with respect to the centerline of the pipe. Two fluids in Figure 23.2 behave like pseudoplastic fluids, and one fluid behaves like a dilatant fluid. All three fluids are compared to a Newtonian fluid for $n = 1$. For small values of n, the velocity profiles demonstrate a more square behavior. For $n > 1$, the velocity profiles become more triangular.

For a fully developed laminar flow in a pipe, a force balance on a cylindrical fluid element with radius R and length L results in a relationship between the pressure loss in the pipe and the wall shear stress, namely $-\Delta P = 4 \times \left(\frac{L}{D}\right) \times \tau_{wall}$.

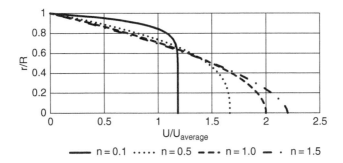

Figure 23.2 Velocity profiles in a fully developed laminar pipe flow with different flow behavior indexes

Now the friction factor for non-Newtonian fluids that follow the power law can be obtained by using the above pressure loss and wall shear stress relationship along with Eq. (23.5). The friction factor for non-Newtonian fluids that follow the power law is given by:

$$f \equiv \frac{\tau_{wall}}{0.5 \times \rho \times \overline{U}^2} = \left(\frac{K}{\rho}\right) \times \left(\frac{3 \times n + 1}{n}\right)^n \times 2^{n+1} \times \overline{U}^{n-2} \times D^{-n} \qquad (23.7)$$

For a fully developed laminar flow in a pipe, the friction factor is related to the Reynolds number by $f = \frac{16}{Re_D}$. Using this relationship and Eq. (23.7), a power law Reynolds number for the non-Newtonian fluids law can be expressed as:

$$Re_{power\ law} = \left(\frac{\rho}{K}\right) \times \left(\frac{n}{3 \times n + 1}\right)^n \times 2^{3-n} \times \overline{U}^{2-n} \times D^n \qquad (23.8)$$

Using the above equations, we will analyze the flow of mayonnaise in a 100 m-long pipe under fully developed laminar conditions at room temperature. Mayonnaise flow in a pipe obeys the power law for non-Newtonian pseudoplastic fluids. Rheometer tests provide us with the following power law relationship for mayonnaise between the wall shear stress and the wall shear rate as shown in Figure 23.3 and between the apparent viscosity and the shear rate as shown in Figure 23.4. The K value for mayonnaise is 70.8 N s$^{0.45}$ m^{-2} and $n = 0.45$. The apparent viscosity of mayonnaise decreases as the wall shear rate increases, as shown in Figure 23.4.

Using the pressure drop relationship to wall shear stress, namely $-\Delta P = 4 \times \left(\frac{L}{D}\right) \times \tau_{wall}$ and Eq. (23.1), we can determine how the pressure drop increases as the pipe diameter decreases, as shown in Figure 23.5 for a 100 m-long pipe. The pressure drop also increases as the power of the flow behavior index (i.e. $n = 0.45$) with increasing shear rate at a constant pipe diameter. The sensitivity of the pressure drop to wall shear stress decreases as the power of the flow behavior index minus one, namely $n = -0.55$, as shown in Figure 23.6. Pressure drops are more sensitive to changes in shear stress at small diameter pipes.

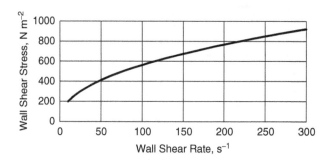

Figure 23.3   Wall shear stress of mayonnaise flow in a pipe versus wall shear rate at 25°C

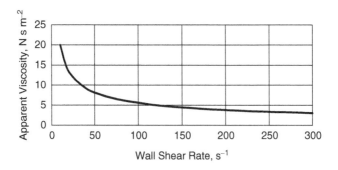

Figure 23.4 Apparent viscosity of mayonnaise versus wall shear rate at 25°C

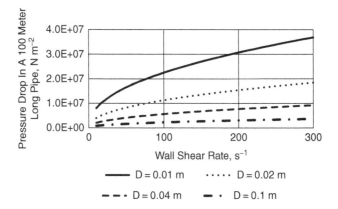

Figure 23.5 Pressure drop versus wall shear rate

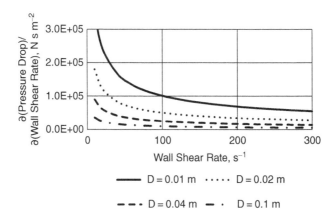

Figure 23.6 Sensitivity of pressure drop to wall shear rate

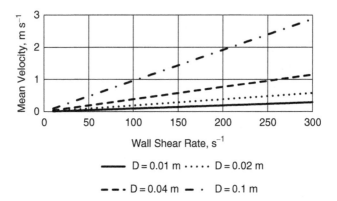

Figure 23.7   Mean velocity versus wall shear rate

The mean velocity can be obtained as a function of the wall shear rate from Eq. (23.5). The mean velocity has a linear relationship with the wall shear rate:

$$\overline{U} = \left(\frac{dU}{dr}\right)_{wall} \times 0.5 \times D \times \left(\frac{n}{3 \times n + 1}\right) \qquad (23.9)$$

The mean velocity increases linearly with the internal pipe diameter and with the wall shear rate. The slope of the mean velocity versus wall shear rate is $0.5 \times D \times \left(\frac{n}{3 \times n + 1}\right)$. The mean velocity versus wall shear rate is shown in Figure 23.7.

The net power required to accomplish a desired volume flow rate for a non-Newtonian fluid that follows the power law relationship between the shear stress and the shear rate can be obtained by the product of the pressure drop in the pipe (i.e. ΔP) and the volume flow rate, namely $0.25 \times \pi \times D^2 \times \overline{U}$. The net power required is given in Eq. (23.10) and the net power required versus the wall shear rate plots are shown in Figure 23.8:

$$Net \ Power \ Required = 0.5 \times \pi \times L \times K \times \left(\frac{n}{3 \times n + 1}\right) \times D^2 \times \left(\frac{dU}{dr}\right)_{wall}^{n+1} \qquad (23.10)$$

In this case, the net power required for a constant diameter pipe increases with increasing wall shear rate to the wall shear rate power of 1.45.

The power law Reynolds number is plotted versus the wall shear rate using Eq. (23.8) and the results are presented in Figure 23.9. The power law Reynolds number increases with increasing wall shear rate to the wall shear rate power of 1.55 or $(2 - n)$. For this analysis, all power law Reynolds numbers are in the laminar flow region. The highest power law Reynolds number for a wall shear rate of $300 \, s^{-1}$ and a pipe diameter of 0.1 m is 65.1.

Now let us investigate a special case, namely a more restricted behavior of mayonnaise flow as a non-Newtonian and pseudoplastic fluid. The flow consistency

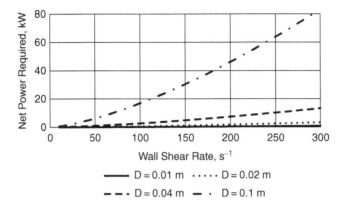

Figure 23.8   Power required versus wall shear rate

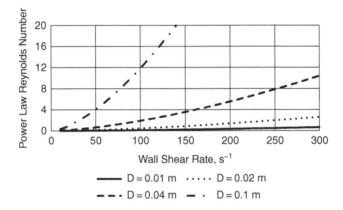

Figure 23.9   Power law Reynolds number versus wall shear rate

coefficient is still 70.8 N $s^{0.45}$ m^{-2} and the flow behavior index is still $n = 0.45$. However, for best preserved taste and long shelf life, we would like the wall shear stress of mayonnaise flow to be $200\,s^{-1}$. The required pipe length in the production plant is taken as 100 m. Also, it is assumed that all pipes have smooth interior walls. The pressure drop versus pipe internal diameter for such a requirement is shown in Figure 23.10. The pressure drop in the pipe behaves as D^{-1}. The sensitivity of the pressure drop to the pipe internal diameter is shown in Figure 23.11. This sensitivity behaves as $-D^{-2}$.

The volume flow rate versus pipe internal diameter for this special case is shown in Figure 23.12. The volume flow rate increases with increasing pipe internal diameter as D^3. The sensitivity of the volume flow rate with respect to increasing pipe internal diameter behaves as D^2.

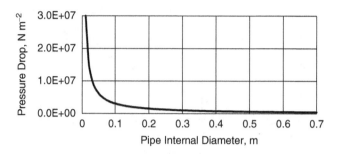

Figure 23.10   Pressure drop in a 100 m-long pipe versus pipe internal diameter for a wall shear rate of 200 s^{-1}

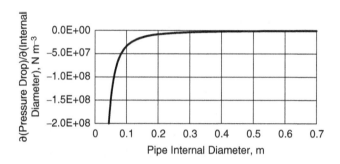

Figure 23.11   Sensitivity of pressure drop to pipe internal diameter for a 100 m-long pipe versus pipe internal diameter for a wall shear rate of 200 s^{-1}

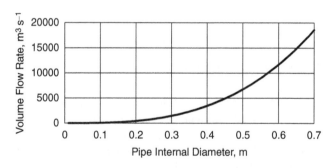

Figure 23.12   Volume flow rate in a 100 m-long pipe versus pipe internal diameter for a wall shear rate of 200 s^{-1}

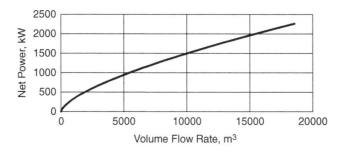

Figure 23.13 Net power required for flow in a 100 m-long pipe versus volume flow rate for a wall shear rate of 200 s^{-1}

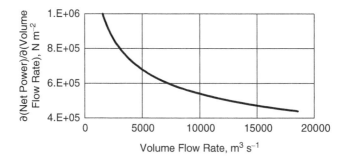

Figure 23.14 Sensitivity of net power required to volume flow rate for a wall shear rate of 200 s^{-1}

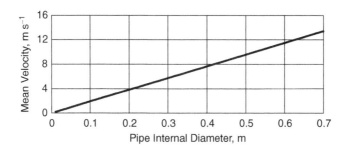

Figure 23.15 Mean velocity versus pipe internal diameter for a wall shear rate of 200 s^{-1}

The net power required versus the volume flow rate for the present special case is shown in Figure 23.13. The net power required increases as the volume flow rate increases. This net power increases with the volume flow rate to the volume flow rate power of 0.667. The sensitivity of the net power required to increasing volume flow rate behaves as the volume flow rate power of −0.333, as shown in Figure 23.14. This sensitivity is more prominent at low volume flow rates.

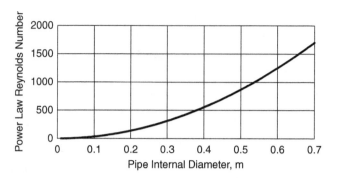

Figure 23.16   Power law Reynolds number versus pipe internal diameter for a wall shear rate of $200\,\mathrm{s}^{-1}$

The mean velocity of mayonnaise flow at a wall shear stress of $200\,\mathrm{s}^{-1}$ for different internal pipe diameters is given in Figure 23.15. This linear relationship has a slope of $19.15\,\mathrm{s}^{-1}$.

The power law Reynolds number for mayonnaise flow at a wall shear stress of $200\,\mathrm{s}^{-1}$ for different internal pipe diameters is given in Figure 23.16. The power law Reynolds number increases as D^2 with increasing internal pipe diameter. All flows in this special case analysis stay in the laminar regime. The maximum power law Reynolds number is 1703 at an internal pipe diameter of $0.7\,\mathrm{m}$.

24

Bubble Rise in a Glass of Beer

In this chapter we will consider a two-phase flow, namely CO_2 bubbles in gas phase flowing upwards in a stationary glass of beer which is in liquid phase. Let us assume that the rising bubble is a rigid sphere and its upward motion is governed by three forces, the upward buoyancy force, downward gravity (bubble weight), and drag forces, as given in Eqs. (24.1)–(24.4):

$$F_{buoyancy} = W_{bubble} + F_{drag} \qquad (24.1)$$

$$F_{buoyancy} = \left(\frac{\pi}{6}\right) \times D^3 \times g \times \rho_{beer} \qquad (24.2)$$

where D (m) is the bubble diameter, constant during its upward travel, g is the gravitational constant, with a value of 9.81 m s^{-2}, and ρ_{beer} is the density of beer, with a value of 1010 kg m^{-3}

$$W_{bubble} = \left(\frac{\pi}{6}\right) \times D^3 \times g \times \rho_{CO2} \qquad (24.3)$$

where ρ_{CO2} is the density of the bubble, with a value of 2 kg m^{-3}

$$F_{drag} = \left(\frac{\pi}{8}\right) \times D^2 \times \rho_{beer} \times U^2 \times C_{drag} \qquad (24.4)$$

where U (m s^{-1}) is the terminal velocity of the rising bubble and C_{drag} is the dimensionless drag coefficient for a rigid and spherical bubble rising in beer (see Ref. [15], chapter 1).

Case Studies in Fluid Mechanics with Sensitivities to Governing Variables, First Edition. M. Kemal Atesmen.
© 2019 John Wiley & Sons Ltd. This Work is a co-publication between John Wiley & Sons Ltd and ASME Press.

Combining Eqs. (24.1)–(24.4), the following expression can be obtained for the terminal velocity of a rising bubble in beer:

$$U = \sqrt{\left(\frac{4}{3}\right) \times \left(\frac{D \times g}{C_{drag}}\right) \times \left[1 - \left(\frac{\rho_{CO2}}{\rho_{beer}}\right)\right]} \tag{24.5}$$

The Navier–Stokes equations for parallel creeping flow past a sphere have been solved by Stokes for cases where the Reynolds number $Re_D = \frac{U \times D}{v}$ is less than 1. For $Re_D < 1$, the drag coefficient was determined to be 24 over the Reynolds number, namely $C_{drag} = \frac{24}{Re_D}$ (see Ref. [15], chapter 6). v (m^2 s^{-1}) is the kinematic viscosity of beer at its bulk temperature. For Reynolds numbers greater than unity, the drag coefficients past a sphere have been determined experimentally. Data correlation to experimental results has been proposed by several scientists. One of those empirical formulas was proposed by Morrison (Ref. [12], chapter 8) and is given by:

$$C_{drag} = \left(\frac{24}{Re_D}\right) + \left[\frac{2.6 \times \left(\frac{Re_D}{5}\right)}{1 + \left(\frac{Re_D}{5}\right)^{1.52}}\right] + \left[\frac{0.411 \times \left(\frac{Re_D}{2.63 \times 10^5}\right)^{-7.94}}{1 + \left(\frac{Re_D}{2.63 \times 10^5}\right)^{-8}}\right]$$

$$+ \left[\frac{0.25 \times \left(\frac{Re_D}{10^6}\right)}{1 + \left(\frac{Re_D}{10^6}\right)}\right] \tag{24.6}$$

Equation (24.6) reduces to the Stokes formula for Reynolds number of unity. By an iterative process the terminal velocity of a rising bubble can be obtained for a given bubble diameter from a combination of Eqs. (24.5) and (24.6) and the Reynolds number. The bubble rise terminal velocity versus bubble diameter is shown in Figure 24.1. The bubble velocity increases almost linearly with respect to

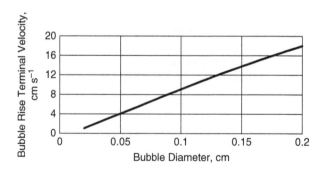

Figure 24.1   Bubble rise terminal velocity versus bubble diameter

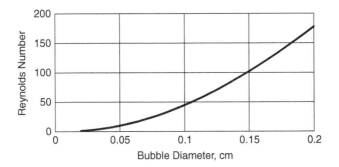

Figure 24.2   Reynolds number versus bubble diameter

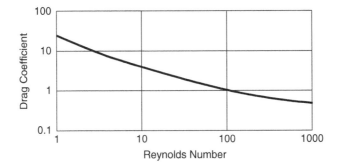

Figure 24.3   Drag coefficient versus Reynolds number in the region of interest, log–log scale

increasing bubble diameter. The slope of a linear curve fit to the line in Figure 24.1 is 95.4 $1\,s^{-1}$.

The Reynolds number versus bubble diameter is presented in Figure 24.2. The Reynolds number increases with increasing bubble diameter as $D^{2.18}$, as a result of a power fit curve to the line in Figure 24.2.

The drag coefficient obtained from Eq. (24.6) is shown in Figure 24.3 in the region of interest on a log–log scale. The drag coefficient decreases as $D^{-1.36}$ with increasing bubble diameter in the region of interest (i.e. $0 < D < 0.2$ cm). If we insert this drag coefficient behavior in the region of interest into Eq. (24.5), we can see that the terminal velocity of the rising bubbles behaves as $U \propto D^{1.18}$, which is very close to a linear relationship.

Let us follow a rising CO_2 bubble from its nucleation site at the bottom of a beer glass to the bottom of the foam layer at the top of the glass. Assume that the initial diameter of the gas bubble at its nucleation site is 0.034 cm, and also that the bubble travels without growing a distance of 20 cm from the bottom of the beer glass to

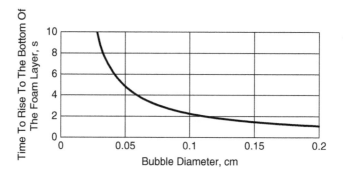

Figure 24.4   Time to rise from the bottom of the beer glass to the bottom of the foam layer at the top of the beer glass versus bubble diameter

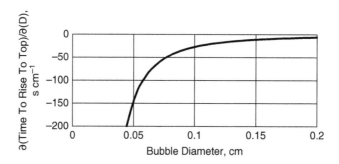

Figure 24.5   Sensitivity of the time to rise from the bottom of the beer glass to the bottom of the foam layer at the top of the beer glass to bubble diameter versus bubble diameter

the bottom of the foam layer at the top of the beer glass. Using Figure 24.1, we can determine that it will take 4.66 s for this size of bubble to rise from its nucleation site at the bottom of the beer glass to the bottom of the foam layer at the top of the beer glass. The time to rise from the bottom of the beer glass to the bottom of the foam layer at the top of the beer glass as a function of bubble diameter is shown in Figure 24.4. As the starting bubble diameter increases, the buoyancy forces acting on the bubble increase and the drag forces decrease, due to increasing Reynolds number; therefore the time to rise decreases. The rise time decreases as $D^{-1.13}$ with increasing bubble diameter. The sensitivity of the time to rise from the bottom of the beer glass to the bottom of the foam layer at the top of the beer glass to bubble diameter versus bubble diameter is shown in Figure 24.5. The absolute value of this sensitivity decreases as the bubble diameter increases as $D^{-2.4}$.

   In the above analysis, we did not consider bubble growth due to convection mass transfer of CO_2 as it rises in the beer glass. Beer bubble growth and its volume expansion have been treated by Zhang and Xu [17]. A bubble's diameter growth rate with respect to time due to convection mass transfer from the CO_2 in the beer through the spherical CO_2 bubble's boundary layer was found to be a constant of about $0.009 \, cm \, s^{-1}$, namely $\frac{dD}{dt} = 0.009 \, cm \, s^{-1}$ [17]. For example, with such a linear bubble growth rate with respect to time, it takes $3.46 \, s$ for a beer bubble that has an initial diameter of $0.034 \, cm$ to travel from the bottom of the beer glass to the bottom of the foam layer at the top of the beer glass. The final diameter of the bubble grows to $0.065 \, cm$ as it reaches the bottom of the foam layer at the top of the beer glass. This bubble's diameter increases 1.9 times from its initial diameter during its rise due to CO_2 mass transfer into the bubble. During its rise, the bubble's volume increases, therefore the buoyancy forces acting on the bubble increase and the drag forces decrease, and it travels to the bottom of the foam layer at the top of the beer glass $1.2 \, s$ faster (namely $4.66 \, s$ rise time for this non-growing bubble versus $3.46 \, s$ rise time when the bubble's volume is growing due to CO_2 mass transfer). The acceleration of this growing CO_2 bubble due to convection mass transfer of CO_2 as it rises in the beer is $0.85 \, cm \, s^{-2}$.

References

1. Atesmen, M.K. (2009). *Everyday Heat Transfer Problems – Sensitivities to Governing Variables*. New York: ASME Press.
2. Carman, P.C. (1939). Permeability of saturated sands, soils and clays. *Journal of Agricultural Science* **29**: 263–273.
3. Darcy, H. (1856). Fontaines Publiques de la Ville de Dijon. Librairie des Corps Imperiaux des Ponts et Chaussees et des Mines.
4. Edwards, B.F., Wilder, J.W., and Scime, E.E. (2001). Dynamics of falling raindrops. *European Journal of Physics* **22**: 113–118.
5. Incropera, F.P. and DeWitt, D.P. (1996). *Fundamentals of Heat and Mass Transfer*, 4e. New York: Wiley.
6. Jones, J.B. and Hawkins, G.A. *Engineering Thermodynamics – An Introductory Textbook*, 1963. New York: Wiley.
7. Kays, W.M. (1966). *Convective Heat and Mass Transfer*. New York: McGraw-Hill.
8. Kozeny, J. (1927). Ueber Kapillare Leitung des Wasser sim Baden. In: *Sitzungsber Akad. Wiss. Wien*, vol. **20**, 271–306.
9. Kreith, F. (1965). *Principles of Heat Transfer*, 2e. Scranton, PA: International Textbook Company.
10. Kuethe, A.M. and Schetzer, J.D. *Foundations of Aerodynamics*, 1964. New York: Wiley.
11. Mei, C.C. (2001). *Stokes Flow Past a Sphere*. Cambridge, MA: MIT.
12. Morrison, A. (2013). *An Introduction to Fluid Mechanics*. New York: Cambridge University Press.
13. Sabersky, R.H. and Acosta, A.J. (1964). *Fluid Flow – A First Course in Fluid Mechanics*. Toronto: Macmillan.
14. Schaschke, C.J. (2016). *Solved Practical Problems in Fluid Mechanics*. Boca Raton, FL: CRC Press.
15. Schlichting, H. (1960). *Boundary Layer Theory*. New York: McGraw-Hill.
16. Shames, I.H. (1962). *Mechanics of Fluids*. New York: McGraw-Hill.
17. Zhang, Y. and Xu, Z. (2008). Fizzics of bubble growth in beer and champagne. *Elements* **4**: 47–49.

Index

Case Studies in Fluid Mechanics with Sensitivities to Governing Variables, First Edition. M. Kemal Atesmen.
© 2019 John Wiley & Sons Ltd. This Work is a co-publication between John Wiley & Sons Ltd and ASME Press.